Conspiracy Theories
in the Time
of Coronavirus

Conspiracy Theories in the Time of Coronavirus

A Philosophical Treatment

RICHARD GREENE AND
RACHEL ROBISON-GREENE

OPEN UNIVERSE
Chicago

To find out more about Open Universe and Carus Books, visit our website at www.carusbooks.com.

Copyright © 2022 by Carus Books

All rights reserved. No part of this publication may be reproduced, stored in a retrieval system, or transmitted, in any form or by any means, electronic, mechanical, photocopying, recording, or otherwise, without the prior written permission of the publisher, Carus Books, 315 Fifth Street, Peru, Illinois 61354.

Printed and bound in the United States of America. Printed on acid-free paper.

Conspiracy Theories in the Time of Coronavirus: A Philosophical Treatment

ISBN: 978-1-63770-006-8

This book is also available as an e-book (978-1-63770-007-5).

Library of Congress Control Number: 2021941782

*For Henry with whom we
most love to conspire!*

Contents

Contents

Thanks

Working on this project has been a pleasure, in no small part because of the many fine folks who have assisted us along the way. In particular, a debt of gratitude is owed to David Ramsay Steele at Carus Books, the Communication Studies and Philosophy Department at Utah State University, and the Department of Political Science and Philosophy at Weber State University. Finally, we'd like to thank those family members, students, friends, and colleagues with whom we've had fruitful and rewarding conversations on various aspects of conspiracy theories as they relate to philosophical themes.

The Current Landscape

At the time of this writing the world is in a horrible state. We are in the middle of a global pandemic that, at the time this book was written, has taken approximately 5.5 million lives with no end currently in sight despite there being a number of vaccines available, each of which has been proven to provide outstanding protection against death. As if that weren't enough, the political division that exists in many countries, including perhaps most notably the United States, presents serious existential threats to their most fundamental political systems and institutions—across the globe, democracy, where it exists, is hanging by a thread. Interestingly, conspiracy theories play a huge role in both of these things.

A New Plague

Deep in a cave somewhere in China, a virus mutated inside of a horseshoe bat. These bats aren't bothered much by coronaviruses. New strains regularly develop inside of them, some of them harmful to other creatures, others perfectly benign.

Enter, a pangolin. Pangolins are unique for two related reasons. First, they are the only mammals with scales. Second, they are the world's most trafficked creature. Pangolins call to mind both anteaters and pill bugs. They have long snouts and tongues that they use to forage for ants, termites, and larvae. When predators are near, pangolins roll up into a ball and their scales make them impenetrable to most hungry carnivores.

Somehow, a pangolin came into contact with the novel coronavirus. Coronaviruses are zoonotic diseases, which means that they can spread from one species of animal to another. This might have been the end of the story. The novel coronavirus may have thrived and then fizzled in bats and pangolins, were it not for the introduction of new predators—human beings. The scales of pangolins are highly valued in traditional medicinal practices. A 1938 article in *Nature* describes some of those practices:

> The animal itself is eaten, but a greater danger arises from the belief that the scales have medicinal value. Fresh scales are never used, but dried scales are roasted, washed, cooked in oil, butter, vinegar, boy's urine, or roasted with earth or oyster-shells, to cure a variety of ills. Amongst these are excessive nervousness and hysterical crying in children, women possessed by devils and ogres, malarial fever and deafness.

It's illegal to sell pangolins, but that doesn't stop it from happening. In fact, the taboo related to the animals confers even more social standing to those who are able to acquire them and imbues the scales with more of a mystical status than they had before. As a result, it's likely that the pangolin which contracted the novel coronavirus from a horseshoe bat was poached and traded on the wildlife black-market.

Scene: A wet market in Wuhan, China. Wet markets exist all over the world. They are places where customers can buy fresh meat and produce. In some wet markets, animals are even slaughtered right in front of the customers. As one might imagine, conditions in these places are not always sanitary. Some wet markets are clean, and one might even think of them as models of local, sustainable food practices. Others, however, are breeding grounds for disease. It is in the latter position that we find our pangolin, contagious with SARS-CoV2. Once this virus spread at the market to human beings, globalization ensured that it steamrolled into a global pandemic. It would go on to kill millions of people, while devastating economies, putting a huge strain on medical resources, filling up hospital beds, shutting down schools, and increasing anxiety levels world-wide.

That is, in all likelihood, what happened. Many Americans believe other explanations altogether.

Fake News and Big Lies

We are undoubtedly living in a golden age of conspiracy theories. Conspiracy theories have nearly always existed, and there have been other golden ages in which conspiracy theories have existed in large numbers while playing key roles in the unfolding of significant events, but we've never seen so many conspiracy theories playing such prominent roles.

A number of factors serve to make this the case.

First, conspiracy theories spread much more easily due to the ways in which the Internet connects all of us together, and they do so in real time (this is particularly true since social media became a significant part of the internet).

Second, conspiracy theories have been weaponized by politicians to discredit their political enemies as well as news sources that would speak out against them. Again, this is nothing new, but it is now happening with greater frequency and at the highest levels of government.

Third, there is a level of distrust of our leaders that is unparalleled in recent centuries. This distrust is not just a distrust of political leaders, although there is plenty of that to go around. There is distrust of business leaders, religious leaders, academics, scientists, the press, and just about any other group that might have wielded some influence at one time or another. In some cases, the distrust is warranted—members of each of the above groups have lied to the public at one time or another, although some groups are more prone to doing so than are other groups. In many cases the distrust is simply not warranted.

Fourth, there are a number of psychological factors at play that predispose certain people to be more readily accepting of conspiracy theories, not the least of which are the feelings of powerlessness and anger that a great number of people feel due to the fact that the pandemic has put a great many lives on hold. We find ourselves in a holding pattern due to circumstances beyond our individual abilities to control (although collectively we have more power than many people realize).

It's distressing, to say the least. Accepting conspiracy theories, at least to the extent that doing so gives us people to blame for our predicaments, allows some people to regain a little bit of that lost power. Finally, there are shameless promoters of conspiracy theories with huge followings and large bullhorns with which to spread their theories. Why would they want to do this? Well, it turns out that there is quite a bit of money to be made in peddling conspiracy theories. These are just a few of the factors that have given rise to this golden age of conspiracy theories. We'll discuss many more in the pages to follow.

It is noteworthy that conspiracy theories are so much a part of daily life now. It was not all that long ago that the majority of people were not all that familiar with the term and even fewer had a keen grasp of the concept; now one seldom goes more than a day or so without hearing about some conspiracy theory or other. Certainly, the change in public awareness of them is partially due to the global pandemic, which has itself spawned a great number of conspiracy theories. Conspiracies theories breed in times of crisis. For example, there are conspiracy theories about the origins of the pandemic, there are conspiracy theories about the production, distribution, and contents of the vaccines, and there are conspiracy theories about the virus itself. (Poor Bill Gates! Most of these theories have attached themselves to him at one time or another.) The change in public awareness also has much to do with the current political climate. The top news stories of the day frequently make reference to conspiracy theories.

Despite the increase in awareness of conspiracy theories, there is still much to be said about them. People have the concept, but it is not clearly defined in the minds of many. Moreover, there are a number of questions pertaining to conspiracy theories that need to be addressed. The main purpose of this book is to highlight and answer many of these questions.

We need, for example, to know just what conspiracy theories are. Some maintain that any theory about a conspiracy counts as a conspiracy theory. We'll argue that this is not the case. What sort of features do conspiracy theories have? In what ways do they change and evolve? What features are unique to conspiracy theories? What is the epistemically responsible attitude to take with respect to conspiracy theories? Should they always be rejected? Is there ever a time when one is justified in accepting a conspiracy theory? Some theorists argue that we should all become conspiracy theorists—or perhaps that we already are all conspiracy theorists. (See, for example, "From Alien Shape-shifting Lizards to the Dodgy Dossier" by

M R.X. Dentith and "Everyone's a Conspiracy Theorist" by Charles Pigden, both listed in the References to this volume.) We'll argue that once conspiracy theories are properly understood, neither of these options are acceptable (even if we all turn out to be persons who accept or embrace some theories about conspiracies).

Conspiracy theories also give rise to questions about the obligations of politicians, the nature of existence, and human psychology. What do we bring to bear on the problems raised by conspiracy theories?

Finally, there are important questions to be addressed pertaining to the ethics of conspiracy theories. In just what ways do conspiracy theories harm us? Who is responsible for the promulgation of conspiracy theories? How bad is it to believe conspiracy theories? And, perhaps most importantly, who is responsible for solving the various problems raised by conspiracy theories? Alas, the solution to these problems of is not going to be a simple one. It's not, for example, a matter of merely educating people that believing conspiracy theories is bad, or having everyone take a critical thinking course at their local college. There is an interesting paradox at play here. Seemingly the more one responds to conspiracy theories, that is the more one refutes, rejects, and debunks conspiracy theories, the more entrenched certain believers of conspiracy theories become. For many, accepting conspiracy theories is an essential part of their identity. Responding to conspiracy theories is tricky business.

We hope that these introductory remarks make it clear that conspiracy theories are generally not to be taken lightly, and especially not in light of our current political predicament and against the backdrop of a deadly ongoing global pandemic. The stakes are quite high and the threat is very real. Ironically, to many it doesn't seem that way. A lot of people think conspiracy theories are mostly fun. In some cases, this is correct.

(We do hope you have fun reading about conspiracy theories in this book!) There is something entertaining about thinking that aliens are being held in Area 51 or that birds are not real or that Jim Morrison is still alive. These, however, as we will see, are the exceptions. In a world where conspiracy theorists are being elected to congress, citizens are storming the capitol, and people are casting votes based on conspiracy theories, we can ill afford to take conspiracy theories with a chuckle or a grain of salt.

How to Read this Book

For those of you reading this book who are, in fact, conspiracy theorists, you'll want to search for the hidden messages that were written just for you. Perhaps you might try taking the first letter of each word and seeing if that amounts to anything. Alternatively, you might try stringing together the first words from each paragraph or the first sentences from each new section to see if either of those reveal the secret messages. If these don't yield results, try reading the odd pages backwards, or using all the hints in *The Da Vinci Code* to point you in the right direction. One of these things is bound to work, and if none do, then keep trying. The truth is out there (or in here) and as no one says, fortune favors the diligent.

For those of you reading this book who are not conspiracy theorists, we can tell you that there are no secret hidden messages in the text. We would tell the conspiracy theorists, but they wouldn't believe us anyway. Still, you will be interested to know a bit about how this book is structured. Each chapter begins with a prominent conspiracy theory (or, in some cases, a handful of related conspiracy theories), which get related to the main themes in the chapter. The idea was to make the book both informational for the person interested in conspiracy theories, and philosophical. It's great to learn about both the philosophy of conspiracy theories, as well as about the conspiracy theories themselves. Enjoy!

Part I

---

*Understanding
Conspiracy Theories*

1
Conspiracy Theories Past and Present

The Case of the Fiddling Emperor

On the night of July 19th (or some say the 18th), 64 A.D., a fire broke out in Rome. It began in the merchant area near Circus Maximus, in the heart of the city, at a shop which contained flammable goods. That night there was a strong wind coming off the river Tiber which caused the fire to spread rapidly.

The blaze initially burned for six days and seven nights and subsequently for another three days when it reignited. Ten of Rome's fourteen districts—more than seventy percent of the city—went up in flames. Hundreds of thousands of people either died or were rendered homeless. These are pretty much the known facts of the case (along with a handful of insignificant details about precisely how the fire spread, and so forth).

So, what happened? How did the fire start? Who, if anyone, was to blame? Despite the fact that no one at the time knew the answer to these questions, there was no shortage of conjecture. The historians Suetonius and Tacitus speculated that the emperor Nero was behind the fire. Suetonius suggested that Nero burned the city simply because he could. His account includes (highly suspect) witness testimony that Nero's agents were seen carrying torches through the streets of Rome. He also reports of Nero singing at a location near the fire at the time. This is the origin of the "Nero fiddled while Rome burned story," which shouldn't be taken too literally, as fiddles had not yet been invented.

Tacitus held that Nero was not in Rome at the time of the fire, but rather, was at his palace in Antium (about fifty miles away), where he ordered the city burned so that he could rebuild it to his own specifications. Nero, on this account, had requested and had been denied approval from the Roman Senate to build a number of palaces, which were to be called "Neropolis" in the heart of the city, including his celebrated Domus Aurea (the Golden House). Gutting a large chunk of real estate nicely facilitated Nero's plan.

Nero, according to Tacitus, blamed the Christians for starting the fire. (Stephen Dando Collins, in his book *The Great Fire of Rome*, argues that Tacitus had Nero blaming Egyptians for the fire, and this was later changed to Christians.) A number of Christians, in fact, admitted to being part of the plot to burn Rome. Of course, there is plenty of reason to believe that their confessions may have been coerced. Still, the confessions served Nero's purpose.

Do we have any reason to favor any of these theories? Tacitus, who was in his early teens at the time of the fire, admits that his account is heavily reliant on hearsay and rumors. Suetonius's account is based on one quoted exchange in which a man is reported as saying to Nero "When I am dead, let the Earth be consumed by fire" to which Nero reportedly replied "No, while I live." And yet, each of these conspiracy theories about the great fire of Rome served its purpose, and each has managed to endure, with some percentage of the population accepting them.

Bridging the Gap

We begin with the story about the Roman fire of 64 A.D. for a couple of reasons. First, as countless reporters, political pundits, and clever creators of internet memes have pointed out, there are striking parallels between the actions (or lack thereof) attributed to Nero and those (again, or lack thereof) attributed to Donald Trump during the first eleven months of the coronavirus pandemic and especially to Trump's actions— mostly golfing and claiming without evidence that the election was rigged—just after the 2020 presidential election was called for Joe Biden (more on the conspiracy theory centered on the rigging of the 2020 presidential election to come). Second, and more importantly for our purposes in this chapter, it highlights the fact that conspiracy theories have been around for a very long time. Throughout that time, they have been used for similar purposes, such as discrediting political rivals, furthering political agendas, shifting focus from one issue to another, deflecting blame, scaring citizens, and to some extent, even fighting culture wars. The list goes on and on.

As was made abundantly clear in the introduction to this book, we are in a sort of "golden age" of conspiracy theories. It is widely held, at least by laypersons and casual observers, that this level of acceptance and proliferation of conspiracy theories is completely unprecedented. While there have never been quite as many conspiracy theories as there are floating around today and they have never been quite as impactful as they currently are (neither of which should come as much of a surprise to anyone given the rate at which information now travels) this is not the first time that conspiracy theories have enjoyed this kind of prominence.

One critical difference between contemporary conspiracy theories and their predecessors is that contemporary theories gain traction and spread *much* more easily. Contemporary conspiracy theories are analogous to wildfires. Any particular bit of wildfire is fairly easy to put out, but they spread extremely quickly, and extinguishing an *entire* wildfire can be quite difficult—containment is challenging but essential.

Conspiracy theories from an earlier era, once established, had considerably more staying power. This is due to the fact that in earlier times fact checking was a much more challenging endeavor. Often, key facts were not available to the general public and, if they were, accessing them wasn't just a matter of searching Google. To continue with the fire analogy, these conspiracy theories were more akin to something like a burning oil field or a tire fire. They didn't spread as rapidly, but they were and are very difficult to extinguish. In these cases, containment is not the problem. It was much more difficult to refute conspiracy theories from bygone eras. So, in some ways the situation has improved, at least in theory. People who have a genuine interest in believing what the best evidence supports and have the skills to make reliable discernments have more information at their fingertips to do so. In other respects, however, the situation has become much worse because of how quickly bad information can spread on the internet and how disinclined people are (for one reason or another) to exercise best practices when it comes to assessing information.

Another important similarity lies in the sorts of circumstances that might give rise to certain types of conspiracy theories. As Jan-Willem van Prooijen and Karen M. Douglas point out, conspiracy theories thrive in times of societal crisis. Human history, of course, is a virtually never-ending stream of societal crises. They argue that the ways in which people feel during times of societal crises (fear, uncertainty, being out of control, and so on) motivates them to create conspiracy theories in order to "make sense" out of what is occurring. These explanatory conspiracy theories eventually manage to become part of the historical narrative, which affects the way that people actually remember the events. This account is consistent with recent literature on the high degree of confabulation that naturally occurs when people remember significant events—even those that don't rise to the level of societal crisis—as described, for instance, in Chris Weigel's article "Quotidian Confabulations." If we remember many of the big events incorrectly, it stands to reason that when we find ourselves in crisis situations, sometimes without good explanations ready at hand for why we are in such circumstances, we produce theories to account for the circumstances, seemingly out of thin air. We often do this in ways that make us feel more comfortable or in control. It's common to want to identify a scapegoat—someone to blame to avoid the recognition of the sometimes tragic absurdity of our experiences.

Given all of this, it's not surprising that conspiracy theories have tended to thrive throughout history whenever there was uncertainty or upheaval, which is almost always the case somewhere. If we simply examine the major events that have occurred over the past one hundred years or so, we'll find conspiracy theories associated with each, and in many cases there were or are numerous conspiracy theories. For the purposes of illustration, we will consider a handful of examples. Conspiracy theorists have postulated that World War I might have ended much sooner had Allied forces not had an economic incentive to keep it going. The virus that caused the 1918 flu pandemic was thought to be a bioweapon, delivered by way of aspirin pills created by Bayer—a German pharmaceutical company. The Great Depression was allegedly brought about by a powerful New World–type cabal. There is no shortage of conspiracy theories pertaining to World War II. Most notably, conspiracy theorists have denied the Holocaust, claimed that President Franklin Delano Roosevelt knew about the attack on Pearl Harbor in advance, and have contended that Hitler's death was faked. Famously, there are also a number of conspiracy theories pertaining to President Kennedy's assassination (take your pick here: the CIA did it, the FBI did it, the KGB did it, the Mafia did it).

One conspiracy theory, fueled in part by some obvious and some not so obvious hints from the band itself, held that Paul McCartney of The Beatles actually died in 1967. Conspiracy theorists have claimed that the Moon landing didn't occur, and that the photos of the landing were faked. At the time that the Soviet Union fell, a number of conspiracy theories were floated that it was a hoax designed to catch the West off guard. 9/11 was claimed to be an inside job. The very presidency of Barack Obama was challenged on the grounds that he wasn't eligible to be president, because some alleged that he was born in Kenya rather than the United States. Many of the same conspiracy theorists held that he was secretly a Muslim. One theory that just very recently began making its way into public discourse, is that the 2020 California wildfires were started by Jewish space lasers. And finally, the coronavirus pandemic has given rise to numerous conspiracy theories, including, but not limited to the standard pandemic line that it was a lab-created bioweapon, and that it was created so that Bill Gates could produce a vaccine with a tracking chip in order to track people's whereabouts and activities (The irony that this conspiracy theory was most often spread via cell phone—devices that track the whereabouts of people—is not to go unnoticed.) Each of these events to some degree coincided with real upheaval in the lives of many people.

A look at significant events prior to the twentieth century yields similar findings. (The article by Mark R. Cheathem shows that the parallels between some of those events and our most recent presidential elections are eerily similar.) There have been many conspiracies about prominent historical figures. Take, for example, the death of President Abraham Lincoln. Some claim that his assassination was orchestrated by high-ranking Confederate officials such as Confederate President Jefferson Davis and Confederate Secretary of State Judah P. Benjamin (which was a conspiracy theory two-fer, as Benjamin was Jewish, and conspiracy theorists frequently target Jewish persons).

Conspiracy theories about the Bavarian Illuminati remain prominent (the Illuminati really existed, but they were initially just six guys who met for drinks and didn't exactly attempt to take over the world). Some challenge the authenticity of the works of William Shakespeare. One prominent theory is that they were not, in fact, written by Shakespeare, but rather were penned by Christopher Marlowe, who allegedly faked his own death and subsequently wrote under the name William Shakespeare. Another less prominent theory is that they were written by Sir Francis Bacon. There were conspiracy theories about the Black Death. In this case, neither explanation was particularly evidence-based, but the conspiracy theories challenged the generally accepted view of scholars of the day, namely, that the plague was the result of Saturn, Jupiter, and Mars all aligning in very specific region of Aquarius on March 20th, 1345. Instead, conspiracists alleged that witches were to blame. Theories about Jesus allege that he was secretly married to Mary Magdalene. And, of course, there are the theories about Nero and the afore-mentioned Roman fire of 64 A.D.

The dominant events and figures throughout history dramatically impact people's lives and how they view the world and their position. As a result, it's not surprising that theories emerge surrounding these events and figures. This book will be an exploration and analysis of theories of this type and the human in-clinations to believe them.

Lightening Things Up a Bit

While van Prooijen and Douglas are correct to point out that times of societal upheaval give rise to a multitude of conspiracy theories, they are not the sole source of them. For instance, small-scale upheavals have also given rise to numerous conspiracy theories over the years. By "small-scale upheavals", we mean less impactful events that might elicit in individuals the same kinds of emotions (fear, panic, uncertainty, a feeling of being out of control, and so forth), but just not on a widespread scale.

Consider, for example, any number of weather experiments conducted by the United States government. These were interpreted by many as UFO sightings. Of course, the most famous of these is the Roswell incident of 1947. What ultimately turned out to be debris from a balloon launched to spy on Soviet atomic weapons testing was initially thought to be the remains of a flying saucer. The need for an explanation quickly and permanently led to wide-spread belief that the "flying saucer" was an alien vehicle. The conspiracy theorizing didn't end there. The requisite "government cover-up" also included the finding of alien bodies and the occurrence of an alien autopsy. It probably didn't help much that a spokesperson from Roswell Army Airfield initially reported that they had found a "flying disk."

All sorts of "unexplained" phenomena give rise to conspiracy theories. That a seemingly large number of accidents occurred in what has come to be known as the Bermuda Triangle calls for explanation. It turns out that the number of accidents occurring there, statistically speaking, is not abnormal, but once the suggestion was made, conspiracy theories were spawned. Similarly, the lost colony of Roanoke Island—a group of about 115 settlers—disappeared without a trace in the late 1590's. Their disappearance gave rise to a number of theories, such as that they were killed by native persons or possibly Spaniards. In truth, recent archeological research suggests they merely migrated a bit from the island. There's no shortage of individuals looking for some sort of explanation for why their children have autism who have readily accepted conspiracy theories about childhood vaccinations causing it, despite overwhelming evidence to the contrary (along with a retraction by the author of the lone bit of evidence in favor of the thesis). If you have a field with crops in it and you stealthily add a few circles which you photograph from above, conspiracy theories involving aliens or perhaps the supernatural will soon follow (interestingly, the presence of crop circles can almost always be explained by farmers who create them to get a laugh at the folks who spread the conspiracy theories). Again, the list goes on.

It would be misleading to suggest that conspiracy theories are always generated by some sort of upheaval. A relatively recent phenomenon in the grand scheme of conspiracy theories (say, over the last sixty years or so) is what we will term the "fun conspiracy theory." The fun conspiracy theory is a conspiracy theory that spreads in virtue of the fact that folks think it is fun to spread. One of us recalls from his childhood a number of books and movies such as *Chariots of the Gods* (which was both a book and a movie) claiming that many artifacts from the ancient world (such as the Egyptian pyramids, the Moai of Easter Island, Stonehenge, and various sculptures) were developed with the assistance of aliens. These works thrived because people enjoyed the theories, and they enjoyed spreading them.

A more recent example is the "birds aren't real" campaign. This theory begins by "pointing out" that during a recent government shutdown, many people coincidentally reported that they had not seen any birds. Of course, the hearer of this doesn't have any immediate recollection of seeing particular birds during this time (unless they were on a birding trip or working at an aviary, in which case they simply deny the theory), so they are either willing to accept it or at least don't have a strong reason to reject it. The payoff is that since birds only appear when the government is not shut down, what we take to be birds must not really be birds but must instead be fake somehow and used for governmental purposes such as spying on people. Normally the fact that a theory seems implausible and is completely unsubstantiated is a pretty good reason to reject it, but in this case, accepting the theory offers someone the fun that comes with accepting and promoting something far-fetched. We recall hearing about this one from our students who enjoyed endorsing it, even though neither we nor the students were convinced that anyone actually bought it. They dug in for the pure pleasure of maintaining the view.

Another fun conspiracy theory is that Finland isn't real. This theory maintains that the area on maps labelled as Finland is actually not land; rather it is sea that Russia and Japan use for fishing in a way that allows them to exceed fishing limits. In these kinds of cases, since the theory is implausible and nothing much is at stake, it can be amusing to suspend one's disbelief.

One conspiracy theory that seems to be all in fun surrounds the recent devastating snowstorms that knocked out power and water in Texas. According to this theory, which gained a lot of traction on TikTok, the snow in Texas was fake, and sent there by the government using alien technology. Their "evidence" was that the snow from the most recent storm didn't melt when exposed to a flame. The reason, of course, is that the snow turned directly to vapor. If someone is looking for a fun theory to accept and nothing much turns on it, they don't need much more than a little flimsy evidence. The evidence against the fake snow hypothesis is quite easy to find—scientists have been debunking it all over the internet. Nevertheless, you might think it is harmless for people to be befuddled by this counterintuitive phenomenon.

On the other hand, the "fake snow" conspiracy turned into something sinister for some believers—they built it in to their theories about Bill Gates and the pandemic or into politically motivated conspiratorial views of President Biden. This illustrates that even conspiracy theories that may appear to be harmless can quickly become dangerous.

Where Are We Now?

The message of this chapter thus far has been that 1. conspiracy theories are not new, 2. there are similarities between present conspiracies and those of the past, and 3. some conspiracies are more serious than others. It would be a mistake, however, to conclude that there are not significant differences between certain of the conspiracy theories of today and those of the past. We find ourselves at a point in time where conspiracy theories pose grave dangers to our very way of life. Currently, the conspiracy theories pertaining to the pandemic, the so-called Deep State, the 2020 presidential election, and those promulgated by QAnon have divided citizens of the United States so deeply that our functioning democracy is hanging by a thread. These conspiracy theories were at the root of the January 6th insurrection and have led to conspiracy theorists being elected to prominent positions in state and local governments. Our elected officials have made use of conspiracy theories to fight culture wars and promote agendas, as opposed to merely deflecting and assigning blame (again, think Nero here). Current conspiracy theories are increasingly used to make us dislike and distrust one another. Moreover, the current trends show little sign of easing, much less reversing.

In the chapters that follow we will offer a deeper analysis of just what constitutes a conspiracy theory, discuss the psychological factors that motivate acceptance and promotion of conspiracy theories, examine the epistemology and ethics of conspiracy theories, and consider some remedies to our current predicament. Ideally, greater understanding of the role that conspiracy theories play in our lives will serve to offset some of the unfortunate force that they have over us.

2
Conspiracy and Theory

The Death of President John Fitzgerald Kennedy

On November 22nd 1963 a motorcade containing President John Fitzgerald Kennedy was traveling through Dealey Plaza in Dallas, Texas. At 12:30 P.M. President Kennedy, who was riding in a convertible limousine was struck by two bullets, one hitting him in the back of the head and the other hitting him in the neck. Texas Governor John Connally, in the same car, was also struck by a bullet. Approximately thirty minutes later Kennedy was pronounced dead at Parkland Memorial Hospital.

Immediately following the assassination, a witness to the shooting, Howard Brennan, informed police that he had heard a shot and subsequently witnessed a man firing another shot from the Texas School Book Depository. He offered a description of the man. About an hour later, officers from the Dallas Police department apprehended a man matching the description offered by Brennan inside the Texas Theater. Sadly, the first police officer to confront the assailant, J.D. Tippit, was shot and killed. The man who shot and killed both President Kennedy and Officer Tippit was Lee Harvey Oswald.

Oswald maintained his innocence, and was not brought to trial, as he was shot and killed two days later by Jack Ruby while being transported from the city jail to the county jail. Despite the fact that Oswald was never put on trial, the assassination of President Kennedy was investigated thoroughly by the Warren Commission, which concluded that Lee Harvey Oswald had acted alone in killing President Kennedy.

As we mentioned in the previous chapter, the killing of President Kennedy has given rise to numerous conspiracy theories. Some hold that Oswald had accomplices (famously, that there must have been a second gunman on the Grassy Knoll, based on the angle from which Oswald must have shot and the direction of the bullets striking Kennedy). Others hold that there was a cover-up at the highest levels of government, such that the Warren Commission did not produce unbiased results. These conspiracy theorists hold that the CIA, the FBI, and even Vice President Lyndon B. Johnson were involved in the cover-up, and they offer as explanation for the conspiracy theory motives ranging from anger about Kennedy's handling of the Bay of Pigs to Johnson's desire to become president. There are conspiracy theorists who maintain that given the fact that Oswald lived for a time in Russia, the KGB must have been involved. Here the motive for the killing is placed squarely on the fact that Oswald was a disgruntled ex-marine, who now had Communist connections and sympathies. Despite some pretty compelling evidence to the contrary, these and other conspiracy theories pertaining to the death of President Kennedy have endured in the minds of the public. According to a 2003 ABC news poll, less than one-third of Americans accept the conclusion of the Warren commission's report that Oswald acted alone.

The conspiracy theories surrounding the assassination of President Kennedy suggest a good place to launch into our treatment of the nature of conspiracy theories. They involve a conspiracy, in fact, they involve numerous competing conspiracies, and they provide theories as to both *why* the event happened and regarding *what* happened. Thus, our discussion of the nature of conspiracy theories begins by looking at the very ideas of conspiracies and theories, respectively.

A Suitable Beginning?

A natural way to begin discussing conspiracy theories might be to give an account of what conspiracy theories are. This, of course, is easier said than done (and will happen in due time). One temptation, as our treatment of the assassination of President Kennedy suggests, would be to look to the name "conspiracy theory." Perhaps a conspiracy theory is just any theory about a conspiracy or any theory that postulates the existence of a conspiracy. Supposing for the time being that this is correct, we would still want to know precisely what constitutes a conspiracy and what constitutes a theory. Let's take up these questions in turn, before returning to the deeper questions of whether conspiracy theories are just theories about conspiracies, and, if not, then just what constitutes a conspiracy theory.[1]

On Conspiracies

At least some kinds of conspiracies are a special form of making plans. It will be useful, then to think carefully about what distinguishes conspiracies from other kinds of plans. From the outset, it is worth noting that thinking about conspiracies has a different *quale* (it

---

[1] Spoiler alert! The likelihood of a complete account of conspiracy theories amounting to nothing more than theories about conspiracies is precisely the same as a complete answer to the question "What is philosophy?" amounting to nothing more than "lover of wisdom."

just feels different) from thinking about plans of other types or about plans more generally. For many of us, conspiracies, all things being equal, carry negative connotations, or, in any case, they motivate worry, suspicion, or both. It may be the case that we are unjustified in feeling negatively toward conspiracy theories. This is something we'll consider in more detail later. For now, it's worth thinking for a moment about the nature of plans.

Sometimes, we make plans on our own. For instance, we might determine which route to take to work in the morning, knowing that there will be construction. We may recognize that a given goal that we have will take time and attention to complete, so we figure out how best to complete it, step-by-step. Planning of this sort is a form of means-ends reasoning—a form that no one ever confuses with conspiracy. Of course, we can make plans on our own which involve others, but, at some point, if those plans come to fruition, they will no longer be reasoning occurring in isolation; instead they will involve *coordination*.

Often, plans are social agreements, and these are the type we're concerned with here. Plans of this sort involve agreements between two or more people concerning future action. There's a different, but related sense of the word "plans" that does not require agreements between two or more people. For example, one may make plans to finish college and then apply to law school, or one may have plans to spend a year abroad. The account we're giving is not intended to rule out this sense of the word; rather we want to capture what is distinct about persons making plans who are not conspiring as opposed to those who are conspiring.

Outside of the context of discussion about conspiracies, we tend to think of plans as being good things— they played a significant role in the ability human beings had to survive in harsh climates. Human beings don't have sharp teeth, we don't have claws that can do damage, and we aren't very fast. The fact that we survived this long seems like some kind of mistake or cosmic joke, and yet human beings dominate (or infest, depending on your view) the planet. Some thinkers would point to our capacity to reason to explain our survival. The explanation for the fact that Tarzan, or young Mowgli from *The Jungle Book* are different from the non-human animals that raised them is that they can do things like make tools and start and manage fire for their own purposes. (But there is more evidence all the time that human beings are far from alone in their capacity to make and use tools.)

Humans are adept at making tools because of their cognitive capacities, but this capacity alone wouldn't get us very far. We are social animals, and it is when we use our cognitive abilities in groups that we are at our most effective. This is where making plans comes in. The better we are at working together, the more sophisticated plans we can make. We are creatures that are capable of using language to express conditional propositions about the future. We can use our imaginations to conjure up things that have not yet come to pass and to consider forking possibilities about how things might go. This is how we obtain enough food to feed large groups, build pyramids, create social policies, and (ideally) participate in making living a human life better for each person in a position to live one. So, on the face of it, making plans seems like a pretty good thing. If conspiracies carry a negative connotation, they don't seem to borrow that connotation from the concept of making plans more generally.

Nevertheless, it seems equally clear that conspiracies *are* interactions that involve a type of plan. Those who believe that the Moon landing was a hoax believe that people *planned* an elaborate scheme of misinformation. Those who believe that JFK's assassination wasn't conducted by a lone gunman believe that there was a group of governmental officials that *planned* an assassination, and part of that plan was to make it appear as if the murder was committed by Lee Harvey Oswald. So, though conspiracies are, indeed, plans, they appear to be plans of a unique type.

One thing that may seem to set conspiracies apart from other types of plans is that they're constructed in secret. Other types of plans can be constructed anywhere, but conspiracies require something like clandestine meetings (or clandestine phone calls, or clandestine communications). After all, when we think about political conspiracies, we tend to think about futured-directed activities conceived of behind closed doors by a group of select individuals. Nevertheless, this account can't be right. As we shall see, not all plans concocted in secret are conspiracies, and not all conspiracies are concocted in secret.

Consider the case of a mother and father planning a surprise birthday party for their child. They are making plans for some future action, and they are doing so in secret, but it isn't obvious that the parents are *conspiring* to throw their child a surprise birthday party rather than simply *planning* the party.

Consider also a case in which Jane conspires with Tim, a hitman, to kill her husband. They meet at a crowded bar and they discuss their plans openly. Jane specifies where and when she would like the murder to take place, and Tim lets her know just how much money he'd have to be paid in order to do the dastardly deed. It may be ill advised for the pair to discuss their plans in public, but the fact that they have done so doesn't make their scheme any less of a conspiracy. If they were charged with conspiracy to commit murder, it would be of no help to their defense for them to claim, "Oh, but the plan was made in a public place, therefore there was no conspiracy." Whether a plan is or is not a conspiracy seems to be about more than whether it was concocted in secret.

One thing that distinguishes a conspiracy from a plan that is merely secretive is whether the persons doing the planning would want the results of their planning to become known at some point—or at least, not become known until such time as knowledge of the plan becomes functionally irrelevant. Here we have in mind cases where a group might conspire to overthrow a government. They need the plan to be secretive forever, if the plan doesn't work, but, if the plan does work, it must be secretive until the regime they are attempting to overthrow is out of power, at which time, they may actually enjoy some fame for their efforts.

The mother and father planning the surprise birthday party don't want their plans to remain secretive forever. On the child's birthday, they want the entirety of the plan to be known by all involved. Notice that this also has implications for Jane's and Tim's activities. While they didn't manage to be secretive about their plans, they certainly would not want their plans to be known. So, conspiracies require certain intentions. Coconspirators must be planners who intend that their plans remain secretive either permanently or for a sufficiently long time as to render the secretive nature of the plans unnecessary or irrelevant.

On Theories

Let's turn our attention toward the theory side of conspiracy theories. In its simplest form a theory is an account or explanation for some set of facts. That explanation can be about things that have already happened (for example, "I have a theory that the Baltimore Colts threw the 1969 Superbowl to the New York Jets) or it can be forward-looking (for example, "My theory is that the bride is not going to show up for the wedding and will marry the best man, instead."). While this definition covers most uses of the term "theory," it doesn't capture an important feature of the way that the term gets employed: the amount of epistemic force that a theory ought to have.

Sometimes the term "theory" is used to convey something more akin to a hunch. We have a mutual friend that frequently helps out people in need in our community, but doesn't ever want credit for doing so. Upon hearing that someone in our community was given an anonymous care package or that their utility bills were brought out of arrears, we almost always form the conjecture that these were the actions of this particular friend. We theorize that it was likely Sam who did the good deed. This conjecture is not entirely without evidence (we know a little something about Sam's past behaviors, Sam's helpful disposition, and Sam's means, etc.), but overall, this is, at best, scant evidence for the proposition that Sam was the person who performed the good deed.

At the other extreme, we have scientific theories. While it's popular in certain circles to say things such as "Evolution is just a theory," scientific theories come with a high degree of evidential support—they are not "just" theories in the way that hunches are just theories. The standards required for a claim to be embraced by the scientific community are pretty high: such theories must be well-substantiated, have much by way of explanatory power, be testable, repeatable, and offer predictions about other events in similar circumstances.

One conclusion to draw from these considerations is that theories come with varying degrees of justification. One can have very little justification or evidential support for a theory or a considerable amount (or something in between). This points to the fact that there is something independent about a person's reason for believing something and their actually believing that thing. This is a theme that we will come back to many times in this book.

Putting Conspiracies and Theories Together

Now we're in a position to address our question: are conspiracy theories nothing more than theories about conspiracies? One thing that seems clear is that any instance of a conspiracy theory will require, at minimum, a belief about a conjoined conspiracy and theory. But that is a very different claim than the one we are considering. We want to know whether there can there be an instance of a theory about a conspiracy that does not manage to be an instance of a conspiracy theory (and not the other way around).

Let's begin with a simple case of a mere hunch about a conspiracy. Suppose that someone believes without any evidence whatsoever the recently fired president of the company they work for was in fact fired because they had discovered that members of the board of directors were embezzling funds They believe further that the board members faced prison time if they didn't take action, but if the president were fired, they could cover their tracks. This is a theory about a state of affairs and the state of affairs involves a conspiracy among the members of the board. This, of course, while being a theory about a conspiracy, does not much resemble actual conspiracy theories. The key difference is that there is no reference to the evidence for the conspiracy, nor is there any discussion of the ways in which the evidence to the contrary is explained away by the theory. This is nothing more than an unsubstantiated hunch that references a conspiracy. We certainly don't want to consider anyone who believes a hunch (or merely proffers a hunch without believing it) to count as a conspiracy theorist.

Going in the other direction, a theory about a conspiracy that meets the exacting standards of scientific theories (or appropriately analogous standards in terms of how exacting they are, as conspiracy theories need to be repeatable, predictive, and so on) wouldn't constitute a conspiracy theory either. Even though such a theory would be about a conspiracy, the conclusive or nearly conclusive evidence would elevate its status to something like a known fact. A bit later in the book, once we've given our positive account of just what constitutes a conspiracy theory, we'll consider the question of whether conspiracy theories are the sorts of things that can be known to be true or whether once known, they become something else entirely.

So, a proper conspiracy theory cannot be the sort of thing for which no evidence is presented to the potential believer, as we see in the case of hunches, nor can it be the sort of thing for which the evidence is overwhelmingly conclusive, as we see in the case of scientific theories. Conspiracy theories are theories about conspiracies where evidence is offered in favor of the theory, that evidence is not conclusive, and usually explains away the evidence for competing propositions (especially those competing propositions pertaining to the issue at hand which are most generally accepted). To be clear, the evidence offered for conspiracy theories is often quite weak, even laughably weak much of the time. Nor does the evidence involved in conspiracy theories need to even be real evidence. Consider for example, the 'birds aren't real' conspiracy theory. Evidence for the claim that birds are not real typically amounts to the false assertion that when the government shut down, no one actually saw any birds. So, we are using the term "evidence" pretty loosely here. In this context "evidence" just means something offered as evidence, whether true or substantiated or otherwise.

So where does this leave us? A conspiracy theory must be more than a hunch. It's a story in which some evidence, even if it is completely flimsy evidence, is offered in favor of a thesis that something is the case, and is the result of persons conspiring. Moreover, the evidence can't be so strong as to rule out all competing hypotheses about the thing that is the case, as that would rise to the level of known fact.

While these are interesting facts about conspiracy theories and they serve to rule out one possible definition of the term "conspiracy theory," namely that they are theories about conspiracies, they do not provide an acceptable account of just what constitutes a conspiracy theory. That will be the topic of the next chapter.

3
What Are Conspiracy Theories?

The Illuminati

In Bavaria, between 1776 and 1785, a secret society existed: one that is not so secret any longer. They were known as the "Bavarian Illuminati." They were formed on May Day by professor of law and former Jesuit Adam Weishaupt. Their goal was to replace Christianity with a religion of reason. This was not a merely academic exercise; the Bavarian Illuminati considered Christianity to be instrumental in promoting injustice and superstition, which leads to abuses in people's personal lives, and serves to facilitate the abuse of state power.

During their nine years of existence the Bavarian Illuminati had a fairly modest membership, peaking at around 650 members. (Some accounts put their peak membership around two thousand, but that would include members of affiliated organizations, as well as bona fide members of the Bavarian Illuminati.) Despite their relatively low numbers, they attracted the attention of a number of notable political figures and intellectuals, such as Johann Wolfgang von Goethe and Duke Ernest II of Gotha (*Brittanica*, "Illuminati").

Other than recruiting people and constructing the sort of membership structure that Dungeons and

Dragons players and religious cult members could only dream of (each member got their own secret name, and the ranks one might achieve included novices, minervals, lesser illuminati, Scottish knights, priests, regents, magi, and kings!), the Bavarian Illuminati didn't actually accomplish much, although there was some eighteenth-century style gangland activity— mostly philosophical and political disagreements with Freemasons (who were in some cases affiliated with the Bavarian Illuminati) and Rosicrucians. In the end it amounted to a lot of ritual discussion, secret greetings, and drinking. Eventually the Bavarian Illuminati disbanded; there was much internal conflict, and an edict from the Bavarian Government rendered membership in the organization illegal.

The Bavarian Illuminati wasn't the only group known by the name "Illuminati." There were groups dating back as early as the fifteenth century. They shared only the name "Illuminati," a claim to being enlightened, and perhaps some really silly rituals: handshakes, funny names and titles, fancy garments, and quite a bit of chanting. Despite the fact that they were largely unaffiliated with one another; they are now all commonly conceived of as a single organization: the Illuminati!

"The Illuminati," a term we now take to refer to the set of all Illuminati organizations, constitutes perhaps the most paradigmatic instance of a conspiracy theory. It is sort of a supreme conspiracy theory (a set of facts about a certain set of actors) from which a great many other conspiracy theories arise. We can understand the relation between the Illuminati qua conspiracy theory simpliciter and the conspiracy theories involving the Illuminati as analogous to the relation between Immanuel Kant's Categorical Imperative (Act only on that maxim which you can at the same time, without contradiction, will to be a universal law) and the individual categorical imperatives to which it gives rise (for example, Don't lie, don't steal, help others in need). It's the story about the Illuminati that gives rise to the individual conspiracy theory stories about the Illuminati.

So, what is that story? It mostly derives from the fact that the Bavarian Illuminati was a secret society of influential European men of power with goals of bringing about change. Over time the mythos has them becoming as close to all-powerful as mortal beings can be, and ultimately controlling everything with an eye toward bringing about some vague new world order. The current popular story of the Illuminati not only links the Bavarian Illuminati to its predecessors, it extends its history back to the Roman Empire (thanks Ron Howard and Dan Brown!) and forward to the present day.

What makes the Illuminati one of the best (perhaps the most paradigmatic of all) examples of a conspiracy theory is the fact that their secret nature provides a good way to explain away all perceptions to the contrary. For example, consider the conspiracy theory that the Illuminati was behind Watergate. From all outward appearances, Watergate was the Nixon administration's doing—they would pursue any course of action that would help them to win the 1968 presidential election. It certainly doesn't *appear* that some powerful global cabal orchestrated it (it was Nixon's men doing the dirty work). Notice that nothing in the official story is incompatible with it being the case that the Illuminati had infiltrated the highest levels of government in the United States and had made it look like it was merely the Nixon administration's doing. In fact, that's exactly how it *would appear* if something were done by a secret organization that wanted it to look that way. The hypothesis that a conspiracy was pulled off by the Illuminati actually *requires* that it not appear that the Illuminati were involved.

Watergate is just one among many of the conspiracy theories attributed to the Illuminati. They are also alleged to have been involved in the assassination of President Kennedy, they reportedly plant secret sexual messages and images in Disney films, they are responsible for chemtrails, they manipulate entire economies, they control the media, they faked their own death (the Bavarian Illuminati never actually ended, they just appeared to do so), and so on. The list of conspiracy theories that place the Illuminati at the center of the conspiracy is long and distinguished.

Looking for a Good Definition

Let's turn our attention to the goal we alluded to in the previous chapter: providing a good definition of the term "conspiracy theory." We've already seen that just cramming the definitions of "conspiracy" and "theory" together won't be sufficient. What we are looking for is a good philosophical definition. What this often means to contemporary philosophers is providing an account of the necessary and sufficient conditions for whatever it is that is being defined.

While this approach is often fruitful, the situation is often more complicated than that. Sometimes the complications arise because the concept being defined is a fuzzy, vague, or ambiguous concept (think of words such a "tall" or "nearby"). Other times, complications arise due to different ways in which the concept is employed (for example, consider the different ways scientists and laypersons might use terms such as "creek" or "planet"). Complications might arise because the referent of a term might be dynamic or subject to change ("pop culture" is a great example of this—some of the things that count as pop culture today did not exist a hundred years ago). Offering a philosophical definition of something like knowledge or water might be a relatively straightforward philosophical enterprise, but defining something like conspiracy theory will prove to be much more difficult. (The epistemologists reading this book right now are undoubtedly screaming "What do they mean that defining "knowledge" is straightforward? Haven't they seen the mountains of post-Gettier literature on the analysis of knowledge?" We have. It turns out, however, that the externalist versions of virtue epistemology are completely correct and the Gettier problem is easily dispensed with.)

One reason why defining conspiracy theories is complicated has to do with the fact that the term gets applied in a number of different ways. For example, there are those literalists who, despite our admonitions in Chapter 2, would be inclined to call any theory about a conspiracy a "conspiracy theory." Even those with a more nuanced approach to conspiracy theories might quibble about whether something counts as a conspiracy theory—some folks, for example, maintain that Watergate (not the Illuminati version) counts as a conspiracy theory, others think it does not rise to that level, even though a conspiracy was involved. Ideally, once a satisfactory definition of "conspiracy theory" is produced it will be something that most upon hearing it will be inclined to accept.

In light of this, let's begin with people's intuitions about what counts as a conspiracy theory in non-controversial cases (for example, that the Illuminati controls the media). People tend to have pretty strong intuitive senses of what constitutes a conspiracy theory. We can use our intuitions here as a measuring stick for whether our account is working. We can also use them to determine whether other definitions of a conspiracy theory already proffered should be accepted. If a definition moves too far away from what people intuitively and pretheoretically consider to be a conspiracy, then we will know that something in the definition has most likely failed. (What we have in mind here is what ethicists commonly refer to as the "method of reflective equilibrium" in which an equilibrium is struck between one's pretheoretical intuitions and what one's theory hypothesizes.)

One of the chief projects of this chapter is to give a philosophical account of conspiracy theories that explains and elucidates our ordinary usage of the term. This is not to suggest that everyone, or even all native speakers of English, are using the term "conspiracy theory" in the same way. Nor is it to suggest that everyone, or even all native speakers of English, have the same intuitions about what counts as a conspiracy theory. It is to suggest, however, that there are paradigmatic uses that overlap sufficiently, as to be useful attaining the sort of reflective equilibrium described above. Hence our intuitions about conspiracy theories and common usage of the term "conspiracy theory" will play a central role in our doing so.

One worry for our program of gaining a sort of reflective equilibrium between our ordinary usage of the term 'conspiracy theory' and our theoretical account lies in the fact that most people use the term 'conspiracy theory' to refer to things that are nonsense or farfetched or something that is, shall we say, kooky to believe. On the one hand it is often correct to view conspiracy theories that way, on the other hand, sometimes conspiracy theories are true. Given this tension, it will be important for the definition that we settle on to either capture this fact about conspiracy theories (that common usage suggests they are nutty, and yet still might turn out to be true) or at least not do anything to rule out this incongruity. In other words, we want the term to apply to things that might be foolhardy to accept, but not have that entail their falsity.

Some Current Definitions

Before turning to our positive account of just what constitutes a conspiracy theory, let's examine some of the current thinking on the matter. While there have been a number of accounts offered over the years—too many to canvas here—focusing on a handful in particular should serve our purposes. We want to get a sense of the sorts of things that philosophers think about conspiracy theories, and we want to motivate our own account. The three views we will consider tend to be pretty standard views.

In his essay "From Alien Shape-shifting Lizards to the Dodgy Dossier" M R.X. Dentith defends a version of the view that we rejected in the previous chapter, namely, that conspiracy theories are just theories about conspiracies. Moreover, he defines a conspiracy as having three components: "1. a plan between two or more people, 2. work in secret, and 3. towards some end." One worry, of course, about Dentith's account is that it doesn't rule out cases such as the birthday party case from the previous chapter. Or to provide another example, suppose that I theorize that my bosses are secretly working on an amendment to our college's strategic plan that they plan on unveiling it to the rest of us only after all the details have been worked out. While the plan is currently a secret one, and it involves two or more persons working toward some end, this would hardly jibe with our ordinary usage of the term "conspiracy theory." At this point we are inclined to embrace the components of Dentith's account but recognize that his necessary conditions do not constitute a jointly sufficient set for something being a conspiracy theory.

In his essay "Everyone's a Conspiracy Theorist," Charles Pigden offers an account that is similar to Dentith's in that it references secretive plans, and defines conspiracy theories as theories about conspiracies, but adds a moral element. Pigden's definition of a conspiracy is ". . . a secret plan to influence events by partly secret means," and that plan must be "morally suspect, at least to some people."

This addition appears to be a step in the right direction insofar as it serves to eliminate our two counterexamples to Dentith's account (the birthday party example and the college strategic plan example). It also seems to better jibe with our commonsense usage of the term, as "conspiracy theory" normally has the connotation that something morally suspect is occurring. Still, there appear to be counterexamples to the morally suspect requirement. For many conspiracy theories, there are altruistic versions of what occurs in the conspiracy. Consider, for example, versions of the fluoridation or chemtrails conspiracy theories in which the chemicals secretly given to people are for their own benefit. Flat Earthers need not suppose that the reason for having people believe that the Earth is a sphere has a morally suspect component.

Quassim Cassam and David Coady both argue that conspiracy theories have a contrarian component. According to these views, conspiracy theories offer an account that runs counter to the generally accepted narrative. For example, the "official" and generally accepted story is that in 1969 NASA landed the lunar module on the moon and Neal Armstrong walked on the moon. The Moon Landing conspiracy theory holds that this never happened and the event was faked. We maintain that Cassam and Coady are correct about this, and the contrarian component is part of a correct definition of "conspiracy theory."

According to Coady's view, the contrarian component constitutes essentially the entirety of his account. He states that "a conspiracy theory is an explanation that is contrary to an explanation that has official status at the time and place in question." The problem with this is that any number of explanations may be contrary to an explanation with official status at the time and place in question. Were we to take this definition literally, the views of Copernicus, Galileo, Newton, and Einstein would all be conspiracy theories. Even taking this in its intended sense, and not so literally, it seems the right thing to say is that the contrarian component is at most one necessary condition among many for something being a conspiracy theory.

Cassam, while correctly highlighting a number of necessary conditions for something being a conspiracy theory, is not actually attempting to provide a counter-example-proof definition; rather, his concern is understanding why people proffer and accept conspiracy theories. His account reflects this. He states that "conspiracy theories are first and foremost forms of political propaganda." He points out that this doesn't apply to all conspiracy theories, but to most of the widely discussed theories. We'll pick up this line of thought a little later when we discuss the purposes and uses of conspiracy theories. For now, it is sufficient to conclude that though it is true that conspiracy theories are frequently used as political propaganda, this fact is only a feature of some conspiracy theories and is not germane to defining conspiracy theories as a broader phenomenon.

Landing on a Definition

So, here's the dialectic. To this point, we've seen that conspiracy theories involve two or more parties, making plans, to bring about some result. Moreover, the plan referred to in the conspiracy theory must have a contrarian component, such that the narrative present runs contrary to the "official" story or explanation. Finally, the plan must be secret or covert in such a way that the conspirators would not want revealed (or at least not for a very long time).

This dialectic is nearly complete. The final element has to do with evidence offered in a conspiracy theory. It's not enough to merely claim that the "official" story or explanation is false, and to offer a replacement. One must also explain the replacement theory. In other words, a fully realized conspiracy theory will tell a story about how it is that there is no evidence for the alternative view, or why the evidence in favor of the "official" narrative is false. It's typically a feature of the conspiracy that the conspirators made it look the way it does on the official story. Typically, the lack of evidence for the conspiracy theory is explained away.

With this added to the other conditions, we are in a position to offer our definition of a conspiracy theory.

> (CT) An explanation of an event or set of circumstances, which is counter to the prevailing narrative, and involves a plan between two or more parties operating in a secretive manner, such that they don't desire that their actions be discovered (either ever or not until there can be no salient consequence for them of their actions becoming known). Conspiracy theories involve either a lack of compelling evidence or evidence that very few other people have access to. The evidence that is offered in a conspiracy theory provides an alternative to the prevailing evidence, such that it purports to account for the fact that the prevailing evidence must be wrong.

(CT) provides us with an account of conspiracy theories that distinguishes conspiracy theories as people commonly use the term from any mere theory that postulates a conspiracy, but it also provides a basis for seeing the badness in accepting conspiracy theories or in being a conspiracy theorist. Many theorists point out that if merely believing theories about conspiracies is sufficient for one's being a conspiracy theorist, then we are all conspiracy theorists. The common usage of the term "conspiracy theorist" is generally reserved for those whose belief in theories about conspiracies is in some sense contemptable or reprehensible (or in the case of some of the sillier conspiracy theories, such as birds aren't real, is unwarranted). It is imperative that our account of conspiracy theories accounts for the badness of actually being a conspiracy theorist. We'll go into more detail about the badness of being a conspiracy theorist later, but it's clear that it, at least in part, has to do with the lack of evidence in support of the particular conspiracy.

4
The Various Natures of Conspiracy Theories

QAnon

QAnon, perhaps the most remarkable of all conspiracy theories to date, began in the fall of 2017. Its beginning, however, was fairly inauspicious. An anonymous user on the website 4chan, who signed his or her posts "Q", left a series of messages in which it was claimed that disgraced President Donald J. Trump[1] was leading a secret campaign to thwart the sex trafficking efforts of a global cabal of Satan-worshipping pedophiles which was led by prominent members of the Democratic Party—most notoriously Hillary Clinton and Barack Obama—along with powerful business persons such as George Soros and A-list celebrities such as Oprah Winfrey and Tom Hanks. (Later versions of the conspiracy theory described the cabal as "cannibalistic Satan-worshipping sex-trafficking pedophiles.")

At one point, QAnon supporters believed that the culminating event of all of this conspiracy (which many

---

[1] We consider former President Trump to be disgraced for his role in the January 6th 2021 insurrection on the United States Capitol. We will leave it to readers to decide for themselves whether that term should be applied for other reasons, such as promoting the Big Lie, failing to respond to COVID-19, attempting to discredit the fourth estate, and bragging about grabbing women by their "pussys," and so forth.

believed was predictable by numerology) was that President Trump was going to bring the cabal to justice in what, in circles of believers, came to be known as "The Storm." The leaders of the cabal were to be rounded up, given military tribunals, found guilty of their crimes, and in many cases, executed. Q, it was claimed, had knowledge of the sex trafficking cabal because he or she was a government insider with a high security clearance (Q-level clearance, to be precise, hence the name).

While QAnon meets the conditions detailed in the previous chapter for something to count as a conspiracy theory, namely, that there is a secret plan that runs counter to the prevailing narrative which few people have access to the details of, and so forth, it is certainly not your "run of the mill" conspiracy theory. For one thing, it is incredibly dynamic. The details are forever changing in response to the emerging needs of believers. For instance, when Q in his or her cryptic writings (known as "Q drops") made a number of predictions that failed to bear out, the details of the theory were just altered. The Storm was originally supposed to occur on November 3rd, 2017. When that didn't happen, the date was changed to Presidential Inauguration Day in 2021. When that didn't happen, it was changed to March 4th 2021 (the purportedly "real" date of Trump's second term inauguration day. We all know how that turned out).

While the details of the QAnon conspiracy are subject to change, even more noteworthy is the dynamic, ever-expanding nature of its scope. Most traditional conspiracy theories tend to have fairly static scopes. For example, the Illuminati conspiracy theory might, over time, bring about changes with respect to the number of things that the Illuminati allegedly controls, but the sorts of things that constitute the particulars of the conspiracy theory do not—the activities of the Illuminati are always of the same type. This is not true of the

QAnon conspiracy theory. While it was originally about Trump's fight against the global cabal, it was soon about a variety of different things. Pizzagate was eventually subsumed into QAnon, the Mueller Investigation was worked into the QAnon story (it was an attempt to cover up the activities of the evil cabal), the Big Lie (the conspiracy theory that Joe Biden rigged the 2020 Presidential Election) was subsumed by QAnon, as were a number of claims about the Coronavirus pandemic, and so on. QAnon is no longer a simple story about a single event (or a single type of event). It is essentially the set of conspiracy theories that its proponents happen to believe, which have some points of intersection (Trump will save the day and Democrats do evil things), but otherwise are quite distinct from one another.

A second way in which QAnon is not your typical conspiracy theory lies in the sheer number of people who believe it. Given how absurd the details are, it seems nothing short of miraculous that *anyone* believes it, but oddly enough, QAnon has more true believers than pretty much any conspiracy theory ever. Polling on QAnon has not been very conclusive (reports of its adherents vary greatly), but the number of people in the United States that accept all or part of QAnon could be as high as one quarter of all Americans, and if that number is on the high side, the lower estimates are not far off. (For details on polling about QAnon see *The Conversation*, 3/5/21.) In Chapter 11 we'll take a closer look at some of the psychological factors that make it the case that so many seem willing to accept this absurd and completely debunked story. For now, it is enough to point out that other conspiracy theories have nowhere near this level of acceptance.

Of course, if a conspiracy theory becomes confirmed, and the confirmation becomes widely known, then the level of acceptance will exceed that of QAnon. For example, pretty much everyone accepts that in the 1950s and 1960s the CIA surreptitiously tested LSD and other drugs on citizens as part of the MK-Ultra program. When this was just a conspiracy theory (as opposed to a proven fact), it was not widely accepted.

A final way in which QAnon is unusual is the fervor with which people believe it. Most QAnon supporters don't casually embrace the claims in the way in which one might, say, casually believe that it is raining; rather, they are committed to the "truths" of QAnon with the sort of passion that puts them at great odds with those who don't accept their views, and creates a strong bond, almost a kinship, with those who do. This leads some to consider QAnon to be a cult. (In this book we are ultimately agnostic on whether QAnon is a cult, but if it walks like a Branch Davidian . . .) One consequence of this is that many QAnon supporters have become estranged from non-believing friends and family members, and a large number of families and friend circles have been ripped apart. By contrast, notice that one hears very few, if any, stories about people who have ended relationships with loved ones over the question of whether birds are real or over claims that the Earth is flat. Further evidence of the intensity surrounding QAnon is the sheer number of YouTube channels, blogs, social media pages, and websites devoted to analyzing each Q drop. Since QAnon pages are often taken down for spreading misinformation, the precise number is a bit of a moving target, but it is well into the thousands. A quick Google search for "QAnon banned" leads to a number of pages such as this one: <www.bbc.com/news/technology-55638558>, which demonstrate the range of QAnon's influence.

As we mentioned, the claims of QAnon have been completely debunked one by one. In fact, there is good evidence that shows that after Q was banned from Reddit and 4Chan, the person posting as Q on 8chan and its successor 8kun is none other than 8chan and 8kun administrator Ron Watkins. (The HBO documentary *Q: Into the Storm* nicely details these claims.) This is significant because Ron Watkins is not a government official with Q level security clearance. He's just a guy who, along with his father, owns a chat board. Despite the debunking, failed predictions (which Q responds to by saying that some misinformation for some inexplicable reason is necessary for the mission to succeed), shifting goalposts, and the real Q (not Ron Watkins) not having been heard from since 2019, QAnon is still growing in terms of the number of people who accept it, and is growing in influence. (To be clear, all we mean by "the real Q" is the person who first posted as Q on 4chan in 2017. We've been given no reason to believe that whomever did the posting actually had top-secret government information.) Disturbingly, it has made its way into mainstream politics. Some elected officials openly support QAnon's positions (the best example of this is Marjorie Taylor Greene). Others who are noncommittal, such as Lauren Boebert, openly court the votes of QAnon supporters by at least hinting at being supportive. QAnon's staying power is remarkable.

Theories That Can't be Wrong

The account of conspiracy theories given in the previous chapter captures those features of such theories that are individually necessary and jointly sufficient, meaning anything that satisfies those conditions must be a conspiracy theory and anything that doesn't must not be one. In addition to a basic definition, there is still quite a bit to say about the nature of conspiracy theories. They share features that are widespread or commonplace, though those features need not be present in order for the phenomenon under discussion to count as a conspiracy theory. In other words, the features that we will be discussing in this chapter are such that conspiracy theories frequently have, but are not features that they *must* have.

The first of those features is *unfalsifiability*. Many conspiracy theories are structured in such a way that they can't be proven false. This isn't because the evidence in favor of them is simply unimpeachable or because they are manifestly true using the natural light of reason or any such thing. Instead, they are crafted such that, in principle, no evidence *can* possibly speak against them. Typically, this feature is simply some mechanism in the way the story is told: any fact that appears to contradict the theory is met with a "that's what they want you to believe" type response.

Falsifiability is an important part of the scientific method (at least since Karl Popper introduced the notion). If something is not falsifiable, meaning that there is no logically possible way that it might be proven false, it is considered to be scientifically unacceptable. The falsifiability criterion often gets applied beyond the scope of scientific inquiry. This, we maintain, gives us good reason to never accept any claim that is not falsifiable— especially those, such as the claims of conspiracy theorists, that consider unfalsifiability to be a virtue.

Consider the following case of "satanic conspiracy." Many creationists believe that the Earth is only six thousand years old. They offer by way of support for this belief certain selections from the Bible dealing with the passage of time. Of course, there are those pesky dinosaur bones (among much other scientific evidence) to contend with that suggest that the Earth is much, much older than that. In response, many creationists, some with platforms on YouTube, claim that what appear to be dinosaur bones aged tens or even hundreds of millions of years are really just tricks planted by the Devil to misdirect God's children from their faith. According to some stories it is God, rather than the Devil, who placed the bones there in order to test human faith. Either way, the better the evidence is in support of the dinosaur hypothesis, the better at deceit Satan is, or, according to the alternate story, the more difficult a faith test God has constructed for us. There is, in principle, no way to rule out the creationist explanation. Nothing can falsify it. Anything an advocate for scientific explanation might offer would simply be met with "that's all just part of the great deceit" or "that's all just an element of the test."

Consider an event more closely connected by conspiracists to human-driven conspiracies—the assassination of President Kennedy. The findings of the Warren Commission serve to rule out all the conspiracy theories that maintain that Lee Harvey Oswald didn't act alone. The "clever" conspiracy theorist makes the evidence against their claim part of the story. They might say something like, "the Warren Commission report was put together by those who were in on the conspiracy theory in order to misdirect people away from the truth." The better the conspirators, the more their schemes are designed to make the facts that characterize their conspiracy easily subsumed by a different explanation of events, in this case, the "findings" of the Warren Commission. Or so the conspiracist would claim.

Recall that when it comes to QAnon, part of the "official" story is that Q intentionally puts out false or misleading information, as it is necessary for the followers to have *some* misinformation in order for the mission to succeed. This provides them with an easy go-to answer when challenged—any time that some part of the QAnon conspiracy theory gets debunked, the members of QAnon claim that very feature was intentionally provided misinformation, which somehow further corroborates the theory.

Another example is the conspiracy theory that 9/11 was an inside job. Evidence to the contrary is typically met with the claim that the insiders produced that evidence in order to cover up their part. Any information that may seem to implicate Al-Qaeda in the attack "really" just substantiates the government's involvement. Or so the conspiracist would claim.

Recall that the final sentence of the definition of conspiracy theories offered in the last chapter states "The evidence that is offered in a conspiracy theory provides an alternative to the prevailing evidence, such that it purports to account for the fact that the prevailing evidence must be wrong." One way to account for the fact that "the prevailing evidence must be wrong" is to offer an alternative which cannot be falsified. In that sense, unfalsifiability can be understood as one way of meeting that condition. In the next chapter we'll further evaluate that move as we consider the epistemology of conspiracy theories. For now, it's sufficient to point out that it's a move that frequently gets made.

Of course, not all conspiracy theories have the feature that they can't be falsified. For example, conspiracy theories such as Paul McCartney is dead, are easily falsified (a simple DNA test would do the trick). Thus, this is a move that is *almost* always available to conspiracy theorists.

You Don't Want to Be an Idiot, Do You?

Another feature often found in conspiracy theories, albeit not as ubiquitous as the first feature, is a part of the narrative that suggests that anyone who does not embrace the conspiracy is in some way mentally deficient. Proponents of conspiracy theories like to suggest that a failure to believe their particular conspiracy theory would make one naïve or unaware or a sheep or a fool (QAnon likes to refer to itself as "the great awakening", and believers often admonish disbelievers to "wake up!"). While maneuvers of this type don't actually offer any support for the veracity of the conspiracy theory, they have great persuasive force—nobody wants to be naïve or foolhardy. This strategy works particularly well in areas where people already have some skeptical attitudes toward the sources of the standard story. For example, people tend to be pretty dubious of the things that politicians claim. It's fairly easy to get people to believe things such as the claim that the Warren Commission report is false, because they have a healthy distrust of politicians. Anyone that has ever been duped or fooled by a politician is more apt to believe that other things said by politicians are false. As people become increasingly dubious of the press and academics, the "don't be a fool" maneuver becomes easier to pull.

One might argue that Shakespeare is partly responsible for this. In *Hamlet*, the titular character states "There are more things in heaven and earth, Horatio, than are dreamt of in your philosophy." For centuries, people have been dropping this line on experts of all varieties to undermine their claims of expertise. It's a "You don't know everything" move that people take as justification for concluding that they don't know anything. Of course, if a certain conspiracy theory pertaining to the authorship of those works typically attributed to Shakespeare is correct, then we blame Marlowe instead.

A particularly egregious version of this move has to do with being a "critical thinker." Proponents of many conspiracy theories (perhaps none more so than those who are members of QAnon) like to portray themselves as ideal critical thinkers. Their definition of a good critical thinker is one who questions everything, and anyone who doesn't is a sheep or a rube or a fool. Questioning everything means not merely accepting what you are told, even if what you are being told comes from proper experts such as scientists or professional philosophers. And it certainly means being dubious of what is reported by the press, regardless of the journalistic standards employed in any particular report.

If you were to take a college-level critical thinking course, you would almost certainly be told how much credence you should place in the advice and admonitions of various experts. If the expert in question has the right sort of credentials, and the matter is something that an expert can be reasonably expected to sort out (it is not terribly controversial in their field, and so forth), then accepting a claim on the basis of what an expert says is something that we ought to do. Conspiracy theorists are turning this model of critical thinking on its head. The message they convey is that in order to be a critical thinker you must do the opposite. Only those who reject what the experts have to say can properly be said to be critical thinkers. In short, they claim to be critical thinkers in virtue of the fact that they reject evidence. Needless to say, this is a horrible way to determine which proposition to accept ("Gee, tell me what people who know about X say, and I'll be sure to believe something else!"). That said, it's a very effective strategy.

A final irony about the "be a critical thinker" maneuver is that there are literally millions who praise themselves for being critical thinkers because they only accept things that are being told to them by someone in a chatroom who claims to be a high-ranking government official and about which they know literally nothing else. Q has offered no evidence that he or she is who he or she claims to be. This is just one instance of this general phenomenon of late. The Internet is ablaze with people arguing against things experts have told them, because some rando on Facebook or YouTube has said otherwise.

The More Things Change the More They Stay Insane

In addition to the lack of falsifiability and some heavy psychological pressure to accept their claims, there's one other feature common to many conspiracy theories that we would like to highlight: conspiracy theories tend to be quite dynamic. In fact, we've already seen this in our discussion of QAnon. Recall that details about when The Storm was to occur kept changing as the predicted dates passed without The Storm actually occurring. Also recall that the things that were part of the QAnon story were subject to change. QAnon was once just a theory about President Trump stopping a global cabal of Satan-worshipping cannibalistic pedophiles, and now it also includes claims about the origin of the coronavirus, the Big Lie, Pizzagate, and so on. We saw something similar in our discussion of the Illuminati as the original stories of the Bavarian Illuminati subsumed earlier Illuminati stories and morphed them into a single group which purportedly existed during the Roman empire.

This raises an interesting question about which claims count as being part of a particular conspiracy theory. Suppose, for example, that a majority of QAnon members initially only believed the original QAnon story (the one about President Trump and the global cabal). Suppose further that at some later point in time a considerable number of QAnon members also came to believe the wholly unrelated conspiracy theory that Finland isn't real. It seems that the right thing to conclude is that a number of QAnon members believe two distinct theories. Similarly, if QAnon members all believe only the original QAnon conspiracy theory (as far as belief in conspiracy theories goes), but also believe that water is wet, we wouldn't want to conclude that the story about President Trump and the cabal includes the proposition that water is wet.

The payoff here is that for something to be part of a particular conspiracy theory it's not enough for adherents to merely believe that thing (or for most members to believe that thing); rather, there mut be some narrative connection between the claims that the adherents believe. We've seen that the QAnon conspiracy theory manages to be about a lot of different and somewhat unrelated things, so the narrative connection can't be that the content is necessarily closely related. The narrative connection must be that the adherents take the various parts of the conspiracy theory to be connected. QAnon members seem to view the Big Lie, and the origins of the coronavirus, and the evil cabal of Satan-worshipping pedophiles as part of the same set of things, regardless of how absurd or implausible their ability to do so seems.

5
The Epistemology of Conspiracy Theories

The Big Lie

By nearly all accounts the 2020 United States Presidential Election was the most secure in the nation's history (See the "Joint Statement" of the relevant government agencies.). This claim is borne out by the findings of the US Department of Justice as well as hundreds of state and county groups responsible for election oversight, security, and review.

By Saturday November 7th 2020 most major news sources had projected that Democratic Party candidate Joe Biden would defeat President Trump. Almost immediately, and well before the results were certified by any states, the results of the 2020 presidential election were challenged by the incumbent president and his legal team. Over sixty legal challenges were filed in US courts. The Trump team lost each one (or had them dismissed or withdrew them before they could be dismissed). Despite all the evidence to the contrary, Trump and his supporters still maintain that which has come to be known as the Big Lie: that the 2020 US Presidential Election was stolen, and that Donald Trump was, in fact, the winner. (This isn't the only political event in history to be known as "the Big Lie." See, for example, Hitler's remarks on what he calls a "big lie" in *Mein Kampf*. Or better yet, don't.)

In one important sense, the Big Lie began on November 7th 2020. Right around the time the networks projected that Biden would win the election, Trump tweeted that there would be a press conference at the Four Seasons in Philadelphia. In what has to be a serious contender for the title "Strangest Thing to Happen in American Politics" the President's attorney, former New York City Mayor Rudy Giuliani, held a press conference at Four Seasons Total Landscaping (nestled humorously between a sex shop and a crematorium), during which he claimed that the election had been stolen. Specifically, he asserted that there was a significant amount of voter fraud, enough to disqualify a substantial number of ballots, in the form of deceased persons voting.

In the following weeks the charges of voter fraud were expanded to include claims of fraudulent ballots being dropped off in the middle of the night, and voting machines, provided by Dominion Voting Systems, being rigged to either throw out votes for Trump or to change Trump votes to Biden votes. While no evidence was ever produced to substantiate these claims (other than a very small handful of anecdotal stories, which, in most cases were easily and swiftly debunked), they were continually repeated by President Trump (and are still being repeated at the time of this writing), his team, his supporters, conservative media outlets, and an insane number of Republican elected officials. (The one exception to this is the claims about Dominion Voting Systems, which came to an abrupt stop around the time Dominion began filing high-dollar lawsuits against those who made the allegations.)

In another important sense, the Big Lie began well-before November 7th 2020. Throughout the campaign,

President Trump told crowds and reporters that the only way that he could lose this election is if it were stolen. He began suggesting that the Democrats would attempt to steal the election via fraudulent mail-in ballots. It was an interesting strategy. On the one hand, he did succeed in getting millions of his supporters to accept the Big Lie and, perhaps, priming the pump was helpful in that regard. It put him in the position of saying "See, I told you that would happen." On the other hand, the strategy seems unwise. To suggest that it was going to happen without having any evidence that it would, in fact, happen, and to follow that up with claims that it did happen (again, without evidence), served to make Trump look like someone who was going to claim cheating was going on no matter what. It seemed to harm his credibility in the eyes of at least some of those he was trying to convince.

Think of it this way. Suppose you have a friend who every April 1st predicts the San Francisco Giants will win the World Series the following October. If one year, they manage to do so, and the friend says "I knew it! I predicted it!", a perfectly reasonable response to them would be to say they didn't know it; they always predict it, and their predictions are usually wrong. We don't usually consider people to be knowers if they always believe a certain thing, regardless of whether it is true or they have any compelling evidence for it. Knowers are people who are sensitive to the truth; they arrive at true beliefs through the right sorts of processes. This, combined with the fact that Trump has in the past said the same thing in other defeats (most famously, with respect to the popular vote in the 2016 Presidential election, and after a primary loss to Ted Cruz in that same election), serves to make belief in the Big Lie wholly unjustified.

Since there was no real evidence that the Big Lie was true, and it was unlikely to secure a victory for

Trump in the courts, why might anyone accept it or promote it? For one, it's a bona fide conspiracy theory. It certainly meets the criteria established in Chapter 3: it's a story about a conspiracy done in secret that runs counter to the prevailing narrative, and is supported by virtually no evidence, but has a way of accounting for that fact, etc. As we've seen plenty of people love a good conspiracy theory.

It also provides something to cling to for those who just aren't able to accept the fact that Joe Biden won the election. It's always tough when a candidate that one supports loses. In an instant any hope for some part of a future one had envisioned is gone. Being able to claim that the other side cheated can be quite consoling. Most folks, however, outgrow that sort of response to disappointment around the time puberty hits.

President Trump appears to have gained at least three things from the Big Lie. First of all, his legal defense fund raised over 250 million dollars in under three months. It turns out that only a fraction of that actually went to legal defense (according to the report by Matt Gregory). Second, it has kept him in the news. Much of the political discussion in the news and on social media during Biden's first four months in office has focused on the Big Lie, which politicians accept it, whether Republicans should denounce it, whether the January 6, 2021 insurrection, which was based on it, should be investigated by a congressional commission, and so on. A second benefit of Trump's constantly being in the news, is the publicity all things being equal is to his advantage, should he decide to run for President in 2024. Finally, it has provided Trump with a mechanism by which he can continue to exert control over the Republican party. Acceptance of the Big Lie has become a test of loyalty to Trump. Those Republican politicians who speak out against the Big Lie are met with swift punishment. For example, Congressperson Liz Cheney was ousted from her leadership position—she was the Republican Party Conference Chair—after criticizing Trump for the Big Lie (and his role in the January 6th 2020 insurrection). Other critics can expect to face primary challenges in upcoming elections.

There are a number of prominent Republican elected officials that have at one time or another actually denounced the Big Lie, who are now promoting it. This makes for a somewhat unique situation. We have a conspiracy theory that is being promoted by a number of people who don't actually believe it, but, instead, fear the consequences of expressing their actual belief. We'll further explore conspiracy theories that are in some cases promoted without belief in Chapter 13, but those tend to be conspiracy theories that have a humorous element (for instance that birds aren't real or that Epstein didn't kill himself). The Big Lie is certainly not a joke. Its consequences will likely be felt for several years if not for generations.

So, I Shouldn't Believe That the Moon Landing Was Faked?

One of the lines that we've been pushing in this book is that often there is not much by way of evidence for conspiracy theories (again, if there were the conspiracy theory would simply become the prevailing narrative, and cease to be a true conspiracy theory). Recall that the definition that we settled on in Chapter 3 contains a clause that makes reference to this: "Conspiracy theories involve either a lack of compelling evidence or evidence that very few other people have access to." This provides one reason for thinking that one should rarely accept conspiracy theories. This raises a couple of questions. First, precisely why not? Second, under what circumstances should you accept a conspiracy theory?

To address these questions (and a number of related questions), it will be useful to do a little epistemology. The short answers to the above questions are 1. because you typically don't have positive knowledge with respect to conspiracy theories (although you can often know that a particular conspiracy theory is false), and 2. you should believe a conspiracy theory when you can know that that particular theory is true. So, heading a little further down the rabbit hole, we need to get clearer on what constitutes knowledge. It would be a gross understatement to assert that there is disagreement among contemporary epistemologists as to what constitutes knowledge, but we should be able to say enough about the features of knowledge that epistemologists agree on to meaningfully address our questions about conspiracy theories.

The Traditional Account of Knowledge or: Some Things Plato Shouldn't Cave On

Traditionally knowledge was thought to have three components. This is a tradition that dates back to the works of Plato. (Some elements of the traditional account of knowledge can be found in Plato's *Meno* and some elements are found in his *Theatetus*.) The traditional account of knowledge maintains that you have knowledge when 1. the proposition under consideration is true, 2. you believe the proposition, and 3. you have a sufficient degree of justification for believing the proposition under consideration.

One thing that pretty much all epistemologists agree on is that the traditional account is incorrect—it is well-established that there must be more to knowledge than these three conditions.[1] One thing that most epistemologists agree on, however, is that each of these conditions

---

[1] In what is perhaps the most influential three-page paper in the history of philosophy, Edmund Gettier demonstrated in 1963 that we can have a justified true belief which could turn out not to be an instance of knowledge.

is necessary. It is nearly completely uncontroversial that the truth and belief conditions are necessary for knowledge: one can't know things that are false, and one doesn't know things unless one believes them. Most epistemologists also consider that the justification condition is necessary for knowledge, although what counts as being justified is highly controversial. We will take it that most epistemologists are correct about this as we present our analysis of the epistemology of conspiracy theories. If it turns out that we aren't correct about this, we'll simply point out that the things that typically replace justification in those analyses of knowledge that jettison the justification requirement (for example, a belief's being sensitive to the truth or a belief's tracking the truth) can be used to make precisely the same points.

One way of approaching knowledge that is beyond the scope of this book is Bayesianism. Bayesianism cashes out justification and related concepts in terms of probability. There is excellent article by Brett Coppenger and Joshua Heter, which discusses Bayes's Theorem in terms of conspiracy theories.

While there is much hair-splitting and hand-wringing over precisely how to cash out the justification requirement, epistemologists tend to agree that being in possession of good reasons for holding a belief is required.[2] For most people, good reasons for accepting a conspiracy theory are not available. Recall that conspiracy theorists tend to get people to accept conspiracy theories by referring to evidence that isn't available to most people, or worse, said evidence just doesn't exist. Again, this is why conspiracy theorists tend to resort to the sort of shaming described in Chapter 4 (for instance, "You don't want to be naive, do you?").

---

[2] We realize that this sounds like it might favor an internalist account of justification, we don't mean to give this impression. We are using "being in possession of good reasons" as a place holder for any of the following (and more): having good reasons, possessing evidence, having a belief that is reliably produced, having a virtually produced belief, and so forth.

So while a particular conspiracy theory might turn out to be true (they usually don't, but occasionally it happens), and some folks might believe that conspiracy theory, those folks fail to know that the conspiracy theory is true, because they don't have justification for believing the conspiracy theory. In Chapter 12 we'll take a closer look at the circumstances under which one should accept a proposition. For now, we'll just assert that if one doesn't know that something is the case, then one ought not believe that thing, and that certainly applies to conspiracy theories.

But What About All Those Reasons We Were Given?

We've been suggesting that conspiracy theories tend to be presented without evidence, but that part of what makes something a conspiracy theory is the account that is offered of why the prevailing narrative is false. For example, we have maintained that the Big Lie is offered without evidence, but part of the Big Lie is a bunch of details about voting machines being rigged, votes being smuggled in in the middle of the night, dead people voting, people voting twice. To the extent that these things are offered as reasons for the Big Lie, why don't they count as evidence? We need to resolve this tension between our claim about conspiracy theories generally lacking evidence and conspiracy theories containing details that constitute reasons to accept their claims.

To see how both of these claims can be true simultaneously, we invoke the distinction between justificatory reasons and explanatory reasons. Justificatory reasons, as the name suggests, are reasons that justifies one in accepting a conclusion. Explanatory reasons are reasons that are offered as explanations for actions or events. Suppose, for example, I come home to find my car parked on a part of the driveway that I never park it on. I might offer as an explanatory reason the supposition that my son used my car and parked in a different spot. This is a story that explains a state of affairs. Suppose further that it is equally likely that my wife used my car and left it in the different spot. Under the circumstances I wouldn't be justified in believing that my son put the car there. In other words, I might have an explanatory reason (albeit misguided) for my belief the proposition that my son put my car in the different spot, but I fail to have a justifying reason.

This is what quite frequently occurs in the case of conspiracy theories. The details offered to explain why a particular conspiracy theory is the case (and why the prevailing or "official" story is wrong) count as explanatory reasons for accepting the conspiracy theory, but these explanatory reasons are not justified. One doesn't have good reasons for believing the explanatory reasons are true. Justificatory reasons have a normative component—they don't simply provide reasons why one *could* believe a particular explanation; they provide reasons why one *should* believe it. They are reasons one ought to accept because they lead reliably to true conclusions. The fact that Rudy Giuliani stood in front of Four Seasons Total Landscape and asserted that the election was stolen, doesn't mean that one should believe that was the case.

Are There Alien Brains in Vats at Roswell?

As we saw in the previous chapter, conspiracy theorists don't merely attempt to persuade by offering reasons in favor of their views. We saw that attempts to shame people into accepting conclusions (for example, "You don't want to be one of the sheep, do you?") are frequently employed. Here we want to highlight a couple more methods, which are frequently employed by conspiracy theorists, and designed to get people to accept particular conclusions. These methods fall somewhere in between offering solid reasons for accepting a position and some of the more manipulative, non-rational methods. We shall see that what both of these methods have in common is a sort of perversion of epistemic standards.

The first of these methods is similar to something we sometimes encounter in skeptical arguments. Skeptical arguments (at least in the skeptical tradition that has its origins in Descartes's *Meditations*) cast doubt on things that we ordinarily take ourselves as having good reasons to believe by raising the standards for knowing those things. For example, you might take yourself as having good reason for thinking that you know you are eating a sandwich (assuming that you are, in fact, eating a sandwich at that moment). You believe that you are eating a sandwich, it is true that you are eating a sandwich, and you have evidence for their belief—you can see it, you can taste it, you have memories of making the sandwich.

The skeptic at this point interjects, that you don't, in fact, have knowledge because you can't rule out some well thought out skeptical scenario, such as that you are only dreaming that you are eating a sandwich, or you are a *Matrix*-style brain in a vat, being tricked into thinking you are eating a sandwich. The payoff of this is that the skeptic argues that you don't know the thing that you think you know, unless you are certain of that thing. Since we are almost never certain of anything, we don't know almost anything. This move is trading on the mistaken notion that certainty is required for knowledge. While Descartes maintained that view, almost no subsequent epistemologist has accepted it. The skeptic has perverted the epistemic standards for knowing by making it seem that the justification condition is really a certainty condition. Note that the skeptic is not trying to convince us that the skeptical hypothesis is true. They are just suggesting an inability to disprove it leads to the conclusion that most putative knowledge is not actually possible.

The conspiracy theorist is making a similar move. The prevailing or "official" story typically comes with all sorts of evidence; that's how it got to be the prevailing story. The conspiracy theorist, just like the skeptic, argues that you can't rule out the alternative view (the view being put forward by the conspiracy theorist), so you shouldn't accept the prevailing view. Consider the conspiracy theory that 9/11 was an inside job. We have lots of evidence to the contrary: Al Qaeda has taken credit for it, there is quite a bit of information about how the attackers were trained and about how it was carried out, and so on. The conspiracy theorist has little by way of evidence in favor of their claim, so instead they offer a series of "You can't rule it out" claims designed to get people to disregard their actual evidence. This, as was the case with the skeptical argument, isn't sufficient to establish the conspiracy theorist's positive claim, but it's a start. Once the epistemic standards for knowing have been perverted, the rest is simple. A little shaming or manipulation will finish the job.

The second method is not as disingenuous as the first. In fact, it may involve a good-faith epistemic maneuver. Nevertheless, it still involves a perversion of an epistemic standard. Here, the idea is to employ the main idea behind a theory of justification that has, to almost every epistemologist's satisfaction, been rightly debunked: *coherentism.*

Coherentism is a variation on the traditional account of knowledge, at least as we've presented it here. There are two main coherentist theories: the coherentist theory of justification and the coherentist theory of truth. We are not discussing the coherence theory of truth in this chapter. The coherence theory of justification maintains that a belief is justified to the extent that it coheres with other things that you believe. So, if you believe that it is going to rain tomorrow, you have more justification for that belief if you also believe other things that cohere with it, such as that the weather report is calling for rain, the readings on your home barometer suggest rain, your gimpy leg that always hurts just before it rains is acting up, and so forth.

There is a lot more detail to the coherence theory of justification than we need to touch on here. The salient points are 1. according to the coherence theory, additional beliefs that cohere with your overall set of beliefs are justified and also increase your overall justification for other things you believe, and 2. the coherence theory, while attractive, is considered by most epistemologists to be completely implausible. No one seems to remain a coherence theorist for very long.

We suspect that very few conspiracy theorists have heard of the coherence theory of justification. That said, they do employ a move that seems to trade on the idea behind the coherence theory, namely, that if things you believe are consistent with other things you believe, then you should accept both things. This principle is demonstrably false. If you believe without evidence that the sky is falling and also believe without evidence that there is a secret plot to lower the sky, then according to this principle, you ought to believe both things, when, in fact, in the absence of evidence, you should believe neither. Recall that when initially discussing QAnon, we detailed the original beliefs of the QAnon conspiracy theory (those pertaining to Trump stopping the global cabal of Satan-worshipping pedophiles), and that eventually QAnon was expanded to include both Pizzagate and the Big Lie. This expansion is an instance of that coherentist move. As QAnon members began to accept the claims of these theories that cohere nicely with the original story, the credence that QAnon members placed on all their beliefs increased. They felt that they were uncovering more and more information, which had a way of looking like more and more evidence, even though there was no actual evidence for any of it. Again, we have a perversion of epistemic standards, which leads to the adherents of certain conspiracy theories not only embracing more conspiracy theories, but doing so with even greater epistemic confidence.

Testimony, Conspiracy Theories, and Hume on Miracles

Many onlookers over the last few years have marveled at what appears to be multiple major movements of mass hysteria. Throughout the history of philosophy, many philosophers have considered rationality to be the essence of man (and here, we typically mean "man" as a name for a specific group of powerful males. Many of these same philosophers had dim views of women and the oppressed groups of their day). Aristotle offered such an account, it is discussed at length in the meditations of Marcus Aurelius, and Descartes offered arguments that he was, in his essence, a "thing that thinks." This philosophical history has passed a long with it a *very* favorable view about the rationality of human beings. When we're tested, though, which we often are, it often becomes frighteningly clear that human beings, of all ilks, are nowhere near as committed to the principles of rationality or sound critical thinking practices as we would like to think.

For example, in his *Enquiry Concerning Human Understanding*, David Hume reports a local rumor from a town in Spain conveyed to him, with a healthy amount of skepticism, by a cardinal. The story was about a man who had undergone a rather miraculous recovery from an ailment. As Hume describes it,

> He had been seen, for so long a time, wanting a leg; but recovered that limb by the rubbing of holy oil upon the stump; and the cardinal assures us that he saw him with two legs.

The townsfolk were all ardent believers in the miracle, and it was endorsed by "all the canons of the church." The story spread and was believed on the basis of testimony and was able to pass and be sustained as easily as it was in part because of a shared trust among members of the community. Nevertheless, the cardinal himself gave no credence to the story. Despite the fact that many people were willing to testify to its truth, a story about such an event is just not the kind of thing that has any meaningful likelihood of being true. The cardinal, "therefore concluded, like a just reasoner, that such an evidence carried falsehood upon the very face of it, and that a miracle, supported by any human testimony, was more properly a subject of derision than of argument."

Hume relates other stories, common at the time he was writing, of people offering and accepting accounts of miracles. He argues that to adjudicate these matters, our evidence consists in our set of past observations. Miracles are violations of the laws of nature. When we consider whether we ought to believe in miracles on the basis of testimony, we must weigh our past observations of the workings of the laws of nature against our observations regarding the veracity of testimony. The former will always win. We will always have more evidence to support the idea that the laws of nature will remain constant than we will to support the belief in eyewitness testimony which reports that those laws have been broken. He says, "The knavery and folly of men are such common phenomena, that I should rather believe the most extraordinary events to arise from their concurrence, than admit of so signal a violation of the laws of nature."

Hume is not just reporting historical fact but is also prescient (though he might object to that characterization) when he says, "men, in all ages, have been so much imposed on by ridiculous stories of that kind." More than 250 years later, we're contending with the kind of elaborate conspiracies we've described in this book: such as shape-shifting reptilian overlords, 5G towers that transmit coronavirus, vaccines that implant microchips, and wild accusations of widespread voter fraud sufficient to change the outcome of the election. QAnon spread in much the same way that the story about holy water being used to grow new limbs spread—through testimony.

Hume points out that the practice of coming to know things on the basis of testimony depends on certain enduring features of human nature. He says, "Were not the memory tenacious to a certain degree, had not men commonly an inclination to truth and a principle of probity; were they not sensible to shame, when detected in a falsehood: Were not these, I say, discovered by experience to be qualities, inherent in human nature, we should never repose the least confidence in human testimony. A man delirious, or noted for falsehood and villainy, has no manner of authority with us."

Society couldn't function if we couldn't rely on testimonial evidence. The present political climate elicits feelings of impending existential dread—a sense that truth and meaning are bleeding off the page like amateur watercolor, leaving no visible boundaries. The characteristics that Hume describes are being worn down. We've been told not to rely on our memories; it is unpatriotic to pay too much attention to the past. There are no behaviors that should make anyone feel shame; to suggest that someone ought to feel ashamed for deceiving and misleading is to "cancel" that person. In an environment immersed in "alternative facts", there is no inclination toward truth or "principle of probity." It is little wonder that in this environment people favor the likelihood of the existence of liberal pedophilic cannibals over the likelihood that anthropogenic climate change is occurring.

With the possible exception of the lizard people who can transform into humans, these conspiracy theories aren't violations of the laws of nature. That said, a similar kind of inductive argument is possible. Most of these conspiracy theories require a level of seamless complicity among many, many people, who then leave behind no compelling evidence. Election fraud conspiracies, for example, require complicity across states, political parties, and branches of government. So, we're left with two broad options. Either every person played their role in this flawlessly, leaving behind no trace, or the theory is false, and it arose from "the knavery and folly" of human beings as has so often happened throughout human history. There is a much stronger inductive argument for the latter.

As Hume points out, humans have certain dispositions that incline them toward truth. On the other hand, they also have strong tendencies to believe nonsense, especially if that nonsense is coherent with what they already believed or otherwise makes them feel good. We might say that everyone ought to have higher epistemic standards, but 'ought implies can'—it makes no sense to say that a person ought to use better methods to form their beliefs when their psychologies prevent them from having any control over such things. All of this has a moral component to it, but it is difficult to know exactly how to identify it. We'll turn to these psychological and ethical features of conspiracy theories in upcoming chapters.

6
A Puzzle about Identity

The George Soros Conspiracy Theory

George Soros is a Hungarian-born holocaust survivor, investor, and philanthropist. He is the founder of the Open Society Foundations, through which he has donated over 32 billion dollars of his own personal wealth. The Open Society Foundations support organizations world-wide that promote democracy, human rights, equality, education, justice, health rights, independent journalism, and other "liberal" causes. Soros is a generous person who has devoted his life to making the world a better place.

The George Soros conspiracy theories, of course, paint a much different picture. They began in the early 1990's with antisemitic Hungarian politician Istvan Csurka accusing Israel via George Soros of attempting to take over Hungary (Jewish Telegraphic Agency 1992). Claims of Jewish persons either controlling the world, controlling some significant aspect of it (such as the media), or *attempting* to take control of it, are ubiquitous across many conspiracy theories. This is certainly a recurrent refrain in the conspiracy theories focusing on George Soros. Believers insist that Soros is surreptitiously behind world events and movements. We'll pursue those theories briefly here.

Like many conspiracy theories, the George Soros conspiracy theory begins with a kernel of truth: George Soros does, in fact, offer financial support to liberal causes. This, in combination with early attempts to discredit Soros, made the conspiracy theories involving him virtually inevitable. In the early days of the conspiracy theories, right-wing politicians and media spokespersons attempted to discredit Soros by circulating a picture of a young person in a Nazi uniform, which they claimed was Soros. The picture was accompanied by stories claiming that he was a Nazi collaborator who betrayed other Jewish persons. These stories have been thoroughly debunked, although there are still many who believe them (Reuters 2020). A second wave of attempts to discredit Soros involved talk-show hosts such as Rush Limbaugh and Glenn Beck claiming that he was a puppet master who controlled the Obama White House (Fast Company 2018). Other claims around this time quoted him (falsely) as saying that he wanted to destroy the United States, and alleging that he wants to make all drugs legal. The list goes on.

Once the right had turned this well-meaning philanthropist into some sort of monster, it was easy to put him at the center of conspiracy theories, and that is exactly what has happened since. Soros has been connected to the 2017 Women's March on Washington, Black Lives Matter protests, Colin Kaepernick's kneeling protest (this one comes from the always accurate and astute Tomi Lahren), the activities of Antifa (it is claimed that he owns Antifa, despite the fact that it is not actually an organization), the 2018 Central American migrant caravan, and the protests against Supreme Court nominee Brett Kavanaugh. In some cases, he is said to have funded these things, in others he is claimed to be more active (for example, he is credited with leaving bricks lying around for protesters to throw at the George Floyd marches). Never ones to miss out, QAnon members have co-opted many parts of these stories, considering Soros to be one of the leaders of the global cabal of Satan-worshipping pedophiles, with each of the above activities being in service of the whims and wishes of the cabal. Each of these claims has been refuted as well.

The mechanics of the George Soros conspiracy theory are simple. Since Soros is exceedingly wealthy and is known to donate to various liberal causes, the right merely has to say that George Soros is behind anything they don't like, and their supporters will accept it. No further details or evidence is required ("If Soros is for it, we are against it!"). Ultimately this constitutes a really sad state of affairs. Imagine dedicating your life's work and considerable wealth to making the world a better place, and because it is politically expedient for a group of people to smear your reputation without any evidence of your doing anything wrong, you become a monster in the eyes of millions.

It's also important to note that antisemitism is an *extremely* common element in conspiratorial thinking and always has been. Indeed, it was the propagation of antisemitic conspiracy theories that contributed to making the Nazi terror and death campaign against the Jews possible. As we'll discuss in a later chapter, conspiracy theories frequently involve scapegoating and appeals to people's existing biases. These are two of the bad consequences which frequently result from the promulgation of conspiracy theories—they harm reputations in ways that are not warranted and they elevate existing bias and prejudices.

A Little Personal Identity

In Chapters 3, 4, and 5 we noted that conspiracy theories such as the Illuminati, QAnon, and the Big Lie are prone to change over time. In this chapter, we've seen that George Soros conspiracy theories increase in detail pretty much anytime an event occurs that the political Right in the United States doesn't like. It's puzzling to think that a theory can change with respect to what it postulates or theorizes, and yet somehow still manage to count as the same theory. For example, suppose that one theory maintains that the Earth is at the center of the solar system and another theory maintains that the sun is. It seems pretty clear that these are different theories. This raises an interesting question: what, exactly, makes conspiracy theories such as QAnon, the Big Lie, and the George Soros conspiracy theory the same conspiracy theories as prior versions of themselves?

Before we answer that question, let's take a look at how similar questions (not pertaining to conspiracy theories) get sorted out by philosophers. This branch of philosophy is typically known as metaphysics, and the particular question we're asking specifically has to do with the metaphysical issues pertaining to identity and is concerned with the question: what makes a thing the thing that it is over time, given that things are subject to change. To illustrate, let's begin with an example. A child is born. We can call this child "Rachard." At birth, Rachard is bald, weighs about seven pounds, is about sixteen inches tall, and has blue eyes. Rachard has other properties, but these should be enough to make our point. Fast forward forty years. Rachard now has long brown hair, weighs about 170 pounds, is six feet tall, and still has blue eyes, but they are a different shade of blue. Intuitively pretty much everyone wants to maintain that Rachard is the same person at forty years old that it was when it was as newborn baby, even though nearly every one of its properties has changed. Moreover, none of the actual physical properties that Rachard had at birth have survived. For example, although Rachard had two hands at birth, and still has two hands at age forty, none of the actual cells that existed at birth are present in Rachard's hands now. Each cell has died and been regenerated many times. There is nothing of newborn Rachard that survives, and yet we are inclined to say that forty-year-old Rachard is the same person.

Some philosophers (known as "bundle theorists" or "mereological essentialists") are inclined to solve this puzzle about identity by claiming that Rachard is not the same person at forty as they were as a newborn. On this view, things are nothing more than sets or collections of properties. As soon as one property changes, we have a different set. So Rachard the newborn stopped being the same entity almost immediately after being born. Similarly, bundle theories with respect to conspiracy theories would hold that as the conspiracy theories change, they become different theories.

Some support for the general bundle theorist approach comes from a widely accepted philosophical principle known as "Leibniz's law." Seventeenth century philosopher and mathematician Gottfried Wilhelm Leibniz proposed the following principle, which he called "the Identity of Indiscernibles": *if x and y do not share all the same properties, then x is not identical to y.* While Leibniz proposed this, it makes sense to think of him as offering a precise formulation of a principle that philosophers have been relying on in one form or another since ancient times.

Suppose, for example, that you are to meet Rachard for the first time at a cocktail party. You know that Rachard is six feet tall and will be wearing a brown sweater. You see a seated person wearing a brown sweater. You wonder whether that might be Rachard, but when they stand up, you observe that they are very short (well under six feet tall). Using Leibniz's law, you conclude that since they don't have all the properties you know Rachard to have (i.e., they are not six feet tall), they must not be Rachard. The bundle theorists point out that since newborn Rachard and forty-year-old Rachard do not share all the same properties (in fact, they don't share any properties to speak of), they must not be the same person.

Many philosophers are dubious about the bundle theorist approach. While it easily solves the problem, it yields some pretty counter-intuitive consequences. Consider a case where Rachard steals a car. The police get a description of the thief. They question Rachard. "Did you steal this car?" Rachard, being a good bundle theorist (and a fast thinker), quickly plucks out a hair, changing their properties, and asserts "it wasn't me; the thief had one more hair than I have." To embrace the bundle theory is to hold that one didn't do any of the things one remembers doing, even if it was just moments earlier, since one is changing all the time. Moreover, on this view, things don't actually change (if by change we mean something like remaining the same being, while undergoing or surviving change); rather, things just pop in and out of existence. Finally, as we will see, Leibniz's Law may not be doing the work that it appears to be doing here. In other words, Leibniz's Law may turn out to be compatible with other approaches, as well.

We can reject the bundle theorists' approach to questions of personal identity on the grounds that it is highly counterintuitive. To be clear, the arguments above do not constitute a decisive rejection of the bundle view, but given that our concern is ultimately achieving a sort of reflective equilibrium between our theoretical account of conspiracy theories and our commonsense intuitions about conspiracy theories, its counterintuitive nature renders it useless for our purposes. This is sort of a mixed blessing. On the one hand we lose a quick solution to our puzzle. We can't just make the puzzle go away by asserting that these aren't the same conspiracy theories, after all. On the other hand, we retain the idea that conspiracy theories are dynamic in the way outlined over the previous three chapters.

Some philosophers, such as John Locke, respond to puzzles in personal identity by focusing on features of a person's psychology. This strategy won't work for things that don't have psychologies, such as rocks and plants, but in the case of persons, it is the fact that there is some psychological continuity that makes it the case that they persist through time, even when undergoing significant changes. Forty-year-old Rachard is the same person as newborn Rachard because forty-year-old Rachard has memories going all the way back to early childhood. On some versions of the view, it is not memories doing the work, but, rather, a continuous series of psychological events—there has been some form of consciousness occurring that connects a person through time to earlier versions of themselves. This view gets around the Leibniz's Law problem by maintaining that things, such as Rachard, don't just have properties simpliciter; rather, they have them at particular times. For example, Rachard is a thing that at one time was sixteen inches tall, and at another time was six feet tall. Rachard, on this view was a single thing in virtue of a connected consciousness that had various competing properties at various times.[1]

This strategy is not going to completely solve the puzzle regarding conspiracy theories, as we don't have anything like a shared consciousness or connected set of memories with conspiracy theories. Conspiracy theories are promoted by multiple people at various times, each of whom might have vastly different beliefs about what the theory entails (or about what the explanation for the theory involves). So, we can't connect past iterations of QAnon, the

---

The metaphysics of temporal properties in the literature on personal identity can get pretty tricky. If you'd like to investigate it further, a good place to begin is Haslanger and Kurtz, *Persistence: Contemporary Readings*.

Big Lie, and the George Soros conspiracy theory with current and future iterations of those theories by appealing to something like the psychology of the adherents.

We can, however, appropriate part of this strategy—the bit about Leibniz's law. If for example, we maintain that conspiracy theories change over time, but that their features are temporally individuated, then the trickiest part of the puzzle goes away. For example, we can say that the George Soros conspiracy theory asserted in 2017 that George Soros did a bunch of sneaky things up to that time, including funding the 2017 Women's March on Washington. We can also say that the George Soros conspiracy theory asserted in 2018 that George Soros did a bunch of sneaky things up to that time, including funding the 2017 Women's March on Washington and the 2018 Central American migrant caravan. The theory had different properties in 2018 than it had in 2017, in just the same way that forty-year-old Rachard had different properties from newborn Rachard.

Wittgenstein to the Rescue

The approach just described removes the impediment to considering conspiracy theories that change over time to be the same theories as their earlier iterations, but it doesn't tell in virtue of what they are the same theories. Going back to our previous example, about the Earth being the center of the solar system, we see that there exist constraints on some theory being the same over time on the one hand and its being a completely different theory on the other. Suppose that a single astronomer held the Earth-centric view at one time and the Sun-centric view at another. We would say that her view changed, but not that she holds the same view. We would maintain that her view changed into a different view altogether or that she rejected and replaced her earlier view. So why should we consider some views to undergo change, instead of saying that they get rejected altogether?

Part of our response to this question involves pointing out that not all change involves rejection. If the followers of QAnon were to assert something like "we were wrong, there does not exist a global cabal of Satan-worshipping pedophiles" and replace their belief with "we were duped by the supporters of Donald Trump and the right-wing media, in order to get us to vote against democrats," then the right thing to conclude is that the supporters of QAnon now believe a new theory. Instead, what has happened is that they have revised what they believe, in most cases adding details, such as that George Soros is part of the cabal, and in some cases rejecting things they previously believed, such as that The Storm is coming on January 6th 2020. For the most part, their previous beliefs have remained intact.

Still, that doesn't answer the question: in virtue of what is a theory that asserts a different set of propositions the same theory as an earlier iteration of itself? A further complication arises from the fact that not all persons who subscribe to a particular conspiracy theory at a particular time believe exactly the same things about it. For example, some people believe the Democrats stole the 2020 Presidential election by bringing in fake ballots. Others believe it was stolen using rigged vote counting machines. So, it is tough to pin down exactly what any particular conspiracy theory amounts to unless one focuses on just those things that an individual conspiracy theorist believes. Since no single conspiracy theorist counts as the official spokesperson for the theory, it makes no sense to do that.

Fortunately, there is a nice strategy available for accounting for things that 1. clearly are part of some phenomenon, and need to be accounted for, but 2. escape precise treatment for some reason or another (for example, they involve large collections of beliefs, some of which are not perfectly compatible with some of the others). This strategy comes from Ludwig Wittgenstein.

Allow us to provide just a bit of background. In the early twentieth century, philosophers of language and linguists were concerned with (among other things) providing an account of how language functions. More specifically, they were concerned with addressing the question of how expressions and sentences manage to be meaningful. Wittgenstein, for example, in his early *Tractatus Logico-Philosophicus* argued that sentences are meaningful in virtue of the fact that they constitute pictures of reality. This view came to be known as the "Picture Theory of Meaning." Eventually Wittgenstein came to reject this view. According to a popular story, the veracity of which we cannot vouch for, a colleague of Wittgenstein's with whom Wittgenstein had an acrimonious relationship on one occasion made an obscene gesture toward Wittgenstein (something akin to "flipping the bird"). It occurred to Wittgenstein that this gesture had linguistic meaning, yet was not accounted for by the Picture Theory of Meaning. Rather than replace the Picture Theory of Meaning with some similar theory that attempts to capture in a single thought how language manages to be meaningful, Wittgenstein rejected conceptual analysis altogether.

The Picture Theory of Meaning got replaced by the notion of family resemblance. In his later *Philosophical Investigations* (Remark 66) Wittgenstein argued that language functions in a variety of ways that, while similar to one another, are not reducible to a single function; rather they are "a complicated network of similarities, overlapping and criss-crossing." He provides an analogy between the uses of words and families. Members of a particular family might all resemble one another even though there is no particular characteristic common to all members of that family. So, for example, most of the members might have similar noses, but not everyone has a similar nose, and most of the members might have similar eyes, but not everyone has similar eyes, and so forth. Thus, it may be the case that each member of the family has much in common with each other member, even though there is no single trait or characteristic common to all. Wittgenstein held that the same was true of the ways in which language functions. Moreover, on Wittgenstein's view, we recognize these similarities, just as we recognize that family members look alike, without actually running down the list of traits at a conscious level—we just recognize the resemblance.

Wittgenstein's notion of family resemblance can be employed in capturing what it is that makes a particular conspiracy theory the conspiracy theory that it is. The Big Lie, for example, is the Big Lie in virtue of a number of related things said about the outcome of the 2020 Presidential election, even if some of those things are not compatible. Similarly, the George Soros conspiracy theory is what it is in virtue of the similarities of the actions attributed to Soros, even if there is no single person who believes each of the things attributed to Soros.

So now we have everything we need to answer the various questions raised throughout this chapter. Particular conspiracy theories are collections of beliefs or claims that nicely go together in a family resemblance sort of way, and change over time to reflect those claims that get added to the story. In the event that the new beliefs stand in such a contradiction to the prior beliefs that they no longer have a family resemblance (as was the case with the Earth-centric and Sun-centric theories), then it is appropriate to maintain that the theory has changed into a new theory. As details get added to a particular story, some things stick and others do not.

Similarly, some things remain and others drop out. As long as the current version of the theory bears a sufficiently strong resemblance to earlier versions, it is appropriate to think of it as the same theory.

Part II

*Conspiracy Theories
in the
Modern World*

7
Conspiracy Theories as Jokes

Birds Aren't Real

Birds exist! They are not mythical creatures, although some mythical creatures are birds, such as Aethon, the eagle that famously tormented Prometheus, and some mythical creatures, such as the hippogriff, have bird-like properties.

There are many interesting facts about birds. For example, birds have vertebrae, there are over ten thousand species of birds, they are the only animals on Earth that have feathers, many (most!) of them can fly, some are very small, and others are relatively large, some can swim, some sleep with one eye open, some cover themselves in ants. Many scientists and fans of the *Jurassic Park* franchise of books and movies believe that birds evolved from dinosaurs. Others maintain that birds evolved parallel to dinosaurs. There are many pictures of birds, paintings of birds, stories about birds, and songs about birds. Birds play a huge role in many cultures and have since the earliest cave paintings. So far, so good; none of this is particularly controversial (ignoring for the moment that some people actually deny that evolution occurs).

It is when we assert that birds currently exist in the continental United States that we begin to run into a bit of trouble. According to the Birds Aren't Real conspiracy theory, there are no birds in the continental United States. Rather, between 1959 and 1971, birds in the United States were systematically eliminated by the CIA (they were purportedly poisoned using B52 bombers during a campaign called "Operation Water the Country") and replaced with cleverly designed drones, which are among other things used to spy on US citizens.[1] Some versions of the conspiracy theory have the birds being killed between 1959 and 2001, and still others have the project beginning in the 1970s.

This is the payoff of the Birds Aren't Real conspiracy theory, but the details read like a conspiracy theory greatest hits collection: according to the "official" story the assassination of President Kennedy was enacted by the CIA as retribution for not authorizing the killing of all birds, Area 51 houses a secret bird killing robot development program, the government is using the bird-appearing drones to spy on us, bird droppings are actually government tracking devices, different types of bird drones are used for different purposes (hummingbirds, for example, are used for assassination), and so on. Moreover, the Birds Aren't Real conspiracy theory makes use of pretty standard persuasive devices. For example, part of the "argumentation" in favor of the theory asks people to "Wake up," and to not be naive. There is even a (seemingly to some) plausible way to verify the truth of the theory. Birds Aren't Real conspiracy theorists will say to people, "notice how you don't see any birds during government shutdowns?" presum-

---

[1] For our epistemologist friends, this is Fred Dretske's cleverly designed mule scenario on steroids. With closure with respect to knowledge failing, one might find themself in the position of knowing that what they are seeing is a bird, and not knowing that what they are seeing is not a cleverly designed surveillance drone.

ably that is what you would expect if CIA employees were forced to take those days off from work, as no one would be at work to operate the drones. Of course, people did see birds on those days, they just don't have specific memories of doing so in most cases, so the suggestion will often take root ("Hey, that's right! I don't specifically recall seeing any birds during the government shutdown.").

The Birds Aren't Real conspiracy theory has become extremely popular since its inception in 2017. It trends on social media (thousands follow it on Facebook, Instagram and Tik Tok) and it has become a huge merchandising outfit (always follow the money!), selling T-shirts, hoodies, hats, socks, stickers, and even COVID-19 face masks. You might wonder whether the face masks actually sell well. Many of those most likely to embrace conspiracy theories are also prone to being COVID-deniers. Hence, we wouldn't expect them to be in the market for face masks. Our students bring up the Birds Aren't Real theory repeatedly when we discuss conspiracy theories. It is certainly a favorite in our neck of the woods. That said, despite the fact that literally hundreds of thousands of people claim that birds aren't real, it's not clear that anyone actually believes that it is true.

One fairly ubiquitous argument given as a general reason to never embrace conspiracy theories, unless, of course, the evidence in favor of the conspiracy theory is overwhelming and incontrovertible, trades on the notion that conspiracies require large numbers of people to keep certain details of the conspiracy secret for a very long time. When two or three people are involved, this is quite difficult—details have a way of emerging. When more than two or three people are involved, the likelihood of an event being kept secret plummets rapidly. (There's an interesting account by Matthew Hutson of why people have such difficulty keeping secrets.)

Now consider the vast number of people that would have been involved with a secret CIA program that 1. spent twelve years getting all birds out of the United States, 2. created enough bird-appearing drones to replace all the birds that existed in the United States prior to the secret program, 3. replaced each bird with it's drone doppelgänger, and 4. monitors the data from each drone. There's virtually no way that this could occur in secrecy for what is now going on sixty years. The Birds Aren't Real folks would respond that the secret is out, but notice that none of their evidence is based on leaked information about the program, it's just a story about what happened supported by that silly one liner: "Ever notice how you didn't see any birds during the government shutdown?"

Other Fun Conspiracy Theories

While the more recently instigated Birds Aren't Real conspiracy theory currently sits atop the Mount Olympus of fun conspiracy theories, fun conspiracy theories are by no means a recent phenomenon. Recall our discussion in Chapter 1 of the *Chariots of the Gods* book and movie. These works attributed the existence of the Egyptian pyramids, the Moai of Easter Island, and Stonehenge, among others, to aliens. While the book was published in the 1960s the idea presented therein predate it by centuries. Merely claiming that something was the work of aliens is not sufficient for something being a conspiracy theory, as it lacks the conspiratorial requirement. The stories told about the Egyptian pyramids, the Maoi of Easter Island, and Stonehenge typically include reference to a cover-up of the alien involvement.

What's noteworthy about these conspiracy theories is they don't serve any political purpose. They don't exist to discredit anyone or any group, nor do they exist to promote some particular agenda. They merely satisfy people's desire for fun. People enjoy science fiction, and these conspiracy theories makes science fiction real. Even people who don't ultimately accept them find them fun to read about and think about.

Some "fun" conspiracy theories do exist to discredit others, but not necessarily for the purpose of having others discredited; rather it is for the fun of doing so. In other words, in these cases the person discredited is just being teased, and no one actually believes the details of the discrediting. Here we have in mind conspiracy theories such as the Ted Cruz is the Zodiac Killer conspiracy theory. Clearly Ted Cruz is not the Zodiac Killer, nor does anyone sincerely believe that he was, as he was born a couple of years after the Zodiac Killer began to kill, but people love to suggest that he was. This probably has to do with his stuffy ivy-league public persona and the fact that he is ruthless as a politician. It's a bad combination from a public relations point of view. Whatever the reason, Cruz just seems like an easy target, and people enjoy acting as if the conspiracy theory is true.[2] Ted Cruz is sort of the Johnnie LeMaster of politics. Johnnie LeMaster was a professional baseball player in the 1970s and 1980s. LeMaster played shortstop for the San Francisco Giants. He was a particularly mediocre hitter and fielder, whose name always comes up in the "worst professional baseball player ever" discussions. Early in his career it became a tradition at Candlestick Park to boo him every time he went up to bat or was involved in a play in the field. Even when he did well people booed him. It was considered part of the fun of attending a Giants game. LeMaster was a good sport about it. On one occasion he

---

[1] At this point we would like to go on the record as stating that we do not condone teasing or bullying of any kind (even if the recipient is Ted Cruz). So, we are not, ourselves, suggesting that conspiracy theories of this type are fun, but, rather, that there are those who find them fun.

wore a uniform with the name "Boo" on it. Ted Cruz, by all accounts, is an equally good sport about his teasing, once sending out a tweet on Twitter designed to look like a message from the Zodiac Killer.

A trickier case is the Epstein Didn't Kill Himself conspiracy theory. At least for our purposes, it's tricky because it began as a number of attempts to discredit certain politicians (and for political reasons, as opposed to the Ted Cruz is the Zodiac Killer conspiracy theory), but quickly moved into the "fun" conspiracy theory category.

Convicted sex-offender and extremely wealthy businessperson, Jeffrey Epstein, was found dead in his jail cell on October 19th 2019. Epstein had apparently committed suicide by hanging himself while awaiting trial on charges of sex-trafficking of minors. Almost instantly, conspiracy theories, which suggested that Epstein was murdered began to circulate. Epstein had been a well-connected socialite who hosted lavish parties at his various mansions and private island in the Virgin Islands. It is reported that sex trafficked women were brought to these parties for, shall we say, something other than pleasant conversation. Even worse, there were also allegations that he trafficked underage girls. Both Donald Trump and Bill Clinton were known to have socialized with Epstein. People in Trump's camp suggested that Bill Clinton killed Epstein to hide secrets. People in Clinton's camp suggested similar things about Trump. The hashtags #TrumpBodyCount and #ClintonBodyCount both trended on Twitter.

To this point we have a regular conspiracy theory, or rather a family of conspiracy theories surrounding Epstein's death (Trump and Clinton weren't the only ones who potentially stood to be embarrassed by their connection to Jeffrey Epstein). An event occurred (Epstein's death), we have an official story (the coroners ruling that the death was a suicide), and a number of wholly unsubstantiated claims involving conspiracies that run counter to the official story (Epstein was murdered by guards who were hired to do so). So, what exactly makes this a "fun" conspiracy theory. The answer lies in what happened next. Soon after Epstein's death, vast numbers of internet memes with the expression "Epstein Didn't Kill Himself" began circulating. Even for those who had no interest in the political implications of claiming that Epstein didn't kill himself, it became fun to promote that claim. People began to find clever ways to work it into conversations, such as writing messages where the first lines began with the letters "E," "D," "K," and "H." Finding novel ways to assert that Epstein didn't kill himself became a national activity for a few months. This, like the Birds Aren't Real and Ted Cruz is the Zodiac Killer conspiracy theories, didn't require actual belief for people to report, assent to, and promulgate. So, despite its political origins, for most proponents of the Epstein Didn't Kill Himself conspiracy, it was just a matter of fun.

Two closely related categories of conspiracy theory focus on celebrity death. Some conspiracy theories hold that a certain celebrity faked their own death. The list of celebrities rumored to have done this is long and distinguished. (Many are included in Adam Raymond's list of seventy "great" conspiracy theories.) In some cases, it is suggested that they did it to get out of the limelight or to go back to a simpler lifestyle. Here Elvis Presley and Michael Jackson come to mind. In other cases, it is suggested that they did this to pull the wool over the eyes of a public which is not as clever as them. Perhaps the most famous examples of this are Andy Kaufman and Jim Morrison. Of course, these claims are never backed up with any compelling evidence.

The other category of conspiracy theory that focuses on celebrity death involves claiming that celebrities not believed to have died, are, in fact, dead. The most famous of these cases was the 'Paul McCartney Is Dead' rumors in the 1960s.

It's not clear to us exactly what makes these conspiracy theories fun. But many people do enjoy asserting that some celebrity is really dead, or some celebrity never actually died. Perhaps the fun is related to the general fun of accepting conspiracy theories—being in the know when everyone else is wrong. Perhaps the fun of these conspiracy theories is just related to the fun of being focused on what celebrities are up to (or not up to, as the case may be)—a sort of fun analogous to reading a gossip magazine that one knows to be untrue. At any rate, many people enjoy a good celebrity-based conspiracy theory.

A related category of "fun" conspiracy theory that we will briefly mention here is the So and So Didn't Do the Work Attributed to Them conspiracy theory. Shakespeare, it is suggested, didn't write the plays attributed to him. For a time, it was believed by some that the band Klaatu was really the Beatles. George Lucas is rumored to have directed more of the *Star Wars* movies than are attributed to him.

TheKlaatu example is a really tough nut to crack. Klaatu was a pretty awful band. We get that fans of The Beatles wanted them to get back together, and would gladly accept nearly any iteration of that, but why would anyone want to believe that a band that they revered could put out such lousy music? If we were serious fans of The Beatles and they got back together for real, and the songs they produced sounded just like Klaatu, we would likely start a conspiracy theory that it was Klaatu who was being given credit for the Beatles songs (or, perhaps, we would say that Ted Cruz made those songs).

No Harm in Having a Little Fun, Right?

The above discussion is not meant to be exhaustive; there are certainly many more varieties of "fun" conspiracy theories. These, however, should be sufficient to consider the issue of whether it's bad in any sense to accept fun conspiracy theories. There are at least two ways that come to mind that you might argue for the moral permissibility of "fun" conspiracies. First, you might argue that there's literally no harm in accepting them. Is there really any harm in believing that birds aren't real or that Andy Kaufman is alive or that Epstein didn't kill himself? Second, accepting these sorts of conspiracy theories is fun, and fun has intrinsic value (provided it doesn't come at the expense of others, etc.), especially when it provides a respite or escape from our current political and environmental realities. Both of these ways of arguing are nicely encapsulated in a recent tweet by Kevin Vibert: "Just heard a conspiracy theory that the Loch Ness Monster is actually the ghost of an ancient dinosaur, and since it affects nothing and nobody, I've decided I believe it, as a treat for me."

At this point we're willing to concede that believing "fun" conspiracy theories is probably not the worst thing one could do, and that there is some value to be found in doing so. Still, forming beliefs, even when doing so seems pretty innocuous, as is the case with "fun" conspiracy theories, does not rise to the level of morally permissible. First, accepting things on the basis of insufficient evidence is likely to make us more likely to do so in the future, and this is particularly true with conspiracy theories, where the line between just having fun, and actually believing some pretty reprehensible things can be pretty blurry. Consider again, for example, QAnon. It seems likely that the earliest members of QAnon could not possibly have believed that there existed a global cabal of Satan worshipping pedophiles running a sex-trafficking ring. This must have just seemed like a fun thing to believe about their political enemies. Gradually, the fun element gave way to something really ugly, hateful, and decidedly not fun. It became easier for early QAnon supporters to accept other parts of the theory once they had embraced its origins. Evidence for this claim lies in the ease with which supporters accepted the revised dates of The Storm when previous predictions failed to come to fruition.

Second, acceptance of false beliefs very rarely happens in a vacuum. So, while it is the case that one person might merely be having fun believing that Ted Cruz is the Zodiac Killer or that birds aren't real, others might be influenced to believe these things in a way that is not just for fun. A gullible person who's inclined to distrust government agencies, might easily lose more of their trust when confronted with any conspiracy theories to the effect that the government is engaged in deceptive or untoward practices. We recommend a healthy level of skepticism regarding the admonitions and claims of our elected officials, but a blind rejection of things we are told is counterproductive and harmful. All things considered, it's best if people just believe that birds are real until they've been given good reason to believe otherwise.

8
The Politics of Conspiracy Theories

The September 11th Attacks

On September 11th 2001, four airplanes flying over the United States were hijacked by nineteen members of al Qaeda, an Islamic extremist organization led by Osama Bin Laden. The hijackers crashed two of the jets (American Airlines Flight 11 and United Airlines Flight 175) into the Twin Towers of the World Trade Center in New York City, a third jet (American Airlines Flight 77) was flown into the Pentagon in Washington DC, and the fourth jet (United Airlines Flight 93) crashed into a field not terribly far from Washington DC in Shanksville, Pennsylvania, after the hijackers were overpowered by a group of passengers en route to their (the hijackers') intended destination.

It's not clear what the fourth jet's target was, but some speculate that it might have been either the White House or the US Capitol Building. The Twin Towers, of course, famously collapsed, and the Pentagon sustained serious damage. The death toll from these attacks reached just under 3,000, and included those on the flights, those in the buildings the jets were crashed into, and a significant number of first responders. It is widely regarded as one of the most tragic days in US history.

Shortly after 9/11 Osama Bin Laden took credit for the attacks and not long after that the hijackers were determined to be al Qaeda operatives. Both of these things were further confirmed in the 9/11 Commission Report, which also detailed the plan, the backgrounds of the participants, and the nature of the training the 9/11 hijackers received.

Unlike the Warren Commission Report, the 9/11 Commission Report was intentionally written in such a way as to eliminate, where possible, the sort of ambiguity that gives rise to conspiracy theories. Despite this, 9/11 conspiracy theories are still circulating and have gained a fair amount of acceptance among the general public in the United States. As we've seen with many other conspiracy theories, such as the assassination of President Kennedy, there does not exist a single dominant 9/11 conspiracy theory. Rather, there exist a number of distinct theories, some of which are consistent with most of the known facts, and some of which offer a narrative contrary to the facts as presented above.

Most of the 9/11 conspiracy theories claim that either the 9/11 attacks were carried out by the United States government or the United States Government had advance knowledge of the attacks and allowed them to occur. One reason offered for thinking that the United States Government was involved, either passively or actively, is that the attacks could be used to justify the wars in Iraq and Afghanistan. A second reason offered for thinking the United States Government was involved was financial. People argue that quite a bit of insider trading occurred just prior to the attacks allowing those "in the know" to profit from them and their effects on the various stock and insurance markets.

These "reasons" come in the form of motives for either allowing the attacks to occur with foreknowledge or participating in the planning of the attacks, but they don't constitute actual evidence that the United States government was involved (again, either passively or actively). There is evidence offered, however, and it comes in a number of forms. The most popular 9/11 conspiracy theory maintains that the collapse of the Twin Towers could not have been the result of jet airplanes crashing into them, as that would not have generated enough heat to melt the infrastructure of the towers. Instead, it is argued, the collapse of the Twin Towers must have been the result of a controlled demolition. Presumably, according to those who embrace 9/11 conspiracy theories, this could not have occurred without the government knowing it was going to occur (government employees in the World Trade Center would have necessarily been aware of the demolition efforts). The claim that the crashes were not sufficient to bring about the collapse of the Twin Towers has been refuted by numerous scientists and engineers (for instance, the article by Bažant and Verdure).

Another popular 9/11 conspiracy theory maintains that the Pentagon was not struck by a jet, but rather by a missile fired from a United States military jet. The "evidence" for this claim is the fact that the hole in the pentagon was not as big as an airliner. This, of course, is laughable, as there isn't any reason to believe that the airliner that struck the airplane would have traveled through the exterior wall of the Pentagon completely intact (thus creating a hole in the shape of an airplane). Moreover, actual debris (including the black box!) from flight 77 was found at the site of the crash, and no evidence of a United States missile was retrieved.

In the end, each of the 9/11 conspiracy theories was easily refuted by the volumes of evidence uncovered by forensic scientists and the members of the intelligence community who provided the data that informed the 9/11 Commission Report. But as one would come to expect by now, this has not been sufficient to deter the more steadfast conspiracy theorist.

Conspiracy Theories as Political Weapons

As was pointed out in Chapter 1, conspiracy theories have been used for political purposes since ancient times. Savvy politicians have long recognized that merely sowing the seeds of doubt about a political rival or about some particular event (such as the 9/11 attacks) is sufficient to convince members of a gullible public who are eager to believe whatever it is that they are manipulated into believing.

It did not take much more than mere suggestion to convince those who wanted to believe negative things about former President George W. Bush and former Vice President Dick Cheney that the United States Government was involved in the 9/11 attacks. Similarly, those who have a perpetual axe to grind with the United States government were easily convinced that 9/11 must have been an inside job; manipulation doesn't require that you be aligned politically on one side or the other—you just need to want things to be a certain way. We've witnessed a similar eagerness to accept false claims that come with the flimsiest evidence possible in those who accept the claims of QAnon, the Big Lie, and the George Soros conspiracy theory. Certainly, our natural tendency to engage in confirmation bias along with a strong inclination to perceive the world as we desire it to be is playing a huge role in such matters. Thus, conspiracy theories constitute a political weapon that is nearly always readily available, is easy to use, and comes with a high probability of success.

Perhaps the most common way to weaponize conspiracy theories for political gain is to actually promote the conspiracy theory. This might involve being the one who is providing the "evidence" for the conspiracy theory, connecting the dots, explaining why the standard narrative must be false, detailing the mechanics of the conspiracy, and making sure that the public is aware of the conspiracy theory. This is precisely what former President Trump has done with the Big Lie. As of this writing he is still attempting to delegitimize the Biden Presidency by repeating the Big Lie, offering an account of how it happened (a bit of a moving target this one), and keeping the discussion going in the media.

This way of weaponizing conspiracy theories for political gain is not restricted to those who create the conspiracy theory or those who have something to add to a particular conspiracy theory. One might gain politically just by repeating or continuing to promote a known conspiracy theory. Countless politicians and conservative media personalities have had success achieving their political and/or professional goals by promoting the George Soros conspiracy theory. In some cases, doing so serves to discredit political opponents just for being liberal and in other cases doing so leads to improved television ratings or an increase in the number of books sold.

A second common way of weaponizing conspiracy theories for political gain involves merely publicly accepting the conspiracy theory. This appears to have something in common with the previous example: in both cases one is not the author of the conspiracy theory, but one promotes it. The distinction between this way of weaponizing conspiracy theories for political gain and the one discussed in the previous paragraph, however, is subtle. The political gain currently under discussion doesn't come directly from the broader consequences of the conspiracy being promoted; but rather it comes from being someone who has actually promoted the conspiracy theory.

To see this difference, let's, once again, look at the Big Lie. Acceptance and public endorsement of the Big Lie has become a sort of litmus test for Republican politicians. Those who fail to endorse it risk losing former president Trump's endorsement and, in some cases, risk having him actively tell his supporters to shun them. So, politicians who endorse the big lie gain politically in a way that is distinct from the way that they might normally gain by promoting a particular conspiracy theory (for instance by using the conspiracy theory to convince voters to vote for them). Moreover, in some districts a politician might actually be ultimately harmed by promoting the Big Lie when they get to a general election, but feel that they must embrace it in order to get through a tough primary. We suspect that embracing the Big Lie will be a serious liability in the next Presidential election, but many Republican candidates will need to do so in order to avoid upsetting Trump and his supporters in the Republican primaries.

Similarly, politicians who have repeated the claims of QAnon, such as Marjorie Taylor Greene and Lauren Boebert, do so because there is some political advantage to be had. They are putting themselves in "the same club" as many of the voters in their districts. In cases such as these, endorsing or promoting some particular conspiracy is politically advantageous in the same way that kissing babies, saluting the flag, and loving apple pie is. Fortunately, the number of districts in the United States where embracing the claims of QAnon is politically advantageous are few and scant. Unfortunately, however, that number is rapidly on the rise.

The Harms of Politically Weaponizing Conspiracy Theories

As was just pointed out, the political costs of weaponizing conspiracy theories in the long run can be great. Conspiracy theories tend to get exposed (not that they ever completely go away). The longer the telling of the Big Lie goes on, the fewer people are accepting of it. Politicians who willingly accept conspiracy theories, even if their doing so is in earnest, will eventually have a credibility problem: either they chose to promote lies or they weren't sufficiently diligent in their research. Politicians who promote conspiracy theories do so at their own peril.

Unfortunately, the politicians who weaponize conspiracy theories for political gain aren't the only ones harmed by their doing so. To begin, those persons who are wrongfully implicated in politically motivated conspiracy theories are harmed. Former President Bush is now falsely believed to have either orchestrated or allowed the deaths of three thousand people in the 9/11 attacks. Among those who believe this about him, his reputation is trashed. Moreover, these people don't just passively hold this belief, they actually resent or hate him for it. To the extent that he has desires not to have an unwarranted reputation and to not be despised for things he did not do, he has been harmed by this conspiracy theory. This type of harm doesn't end there. Those on the receiving end of conspiracy theory falsehoods receive death threats, their families are harassed, they sometimes lose their jobs. Any number of bad things can befall someone who is the target of a politically motivated conspiracy theory.

Much the same can be said for corporations or organizations who are wrongly implicated in conspiracy theories. A case in point is Dominion Voting Systems. On one version of the Big Lie it was claimed that Dominion Voting Systems voting machines were manipulated to switch votes for Trump to votes for Biden. Sometimes it was contended that this was done by Dominion employees. Other times it was suggested that the software that Dominion machines used were programmed. These allegations were repeated excessively on conservative media and by President Trump's surrogates. Of course, the claims were never substantiated and were easily debunked.

A second harmful result from the political weaponizing of conspiracy theories is widespread distrust. People now more than ever distrust politicians, political processes, and political institutions. People also more than at any other time distrust experts (see Dr. Fauci!). We contend that this is bad not just for the politicians involved, but for the general citizenry. Many now consider that our very democracy is threatened by the current state of politics, and conspiracy theories such as the Big Lie and QAnon have played a significant role. To the extent that conspiracy theories are leaving large numbers of citizens convinced that our elections can be compromised (or perhaps already have been) we lose the psychological comfort of knowing that our elections will be fair, that the will of the citizenry will be done, and that we need not worry about transitioning to an authoritarian state. One price of this loss of comfort is deep existential dread.

It's the politicians themselves who are doing the most harm to the institutions that they serve. Most professional codes of ethics have some code or other that states that one should not do things that harm the industry or profession. (Codes such as this are often at odds with other codes, such as those requiring whistleblowing under certain circumstances.) Politicians who weaponize conspiracy theories for political purposes would be wise to follow such admonitions. A willingness to discredit political institutions is ultimately a willingness to discredit themselves to some extent.

A third harmful result of politically weaponizing conspiracy theories that has befallen the public is that doing so opens the doors for bad legislation. In the most egregious cases a conspiracy theory, such as the Big Lie, is appealed to as justification for passing legislation to prevent the thing that never happened in the first place from happening in the future. For example, as of July 22nd 2021 eighteen states had passed legislation restricting voting subsequent to the 2020 Presidential election (Brennan Center for Justice). Some of these laws shorten the window for early voting, some restrict voting by mail, some limit which forms of identification one can use when voting. Almost all of these laws make it more difficult for people of color to vote. "Coincidentally" these laws have been passed by Republican legislatures and mostly adversely affect Democratic voters.

A fourth harmful result is that it breeds real anger and resentment among citizens toward one another. Those that accept the Big Lie, the QAnon conspiracy theories, and the George Soros conspiracy theories are angry at those citizens that they believe are either part of the conspiracy or are not sufficiently bothered by it. Those that do not accept the conspiracy are angry at those who do. Our discourse (particularly on social media) has become more hateful and more vitriolic over the past several decades. There are several reasons for this, but chief among them are the roles political conspiracy theories play. One might see them as pouring gasoline on a dumpster fire that is pretty large to begin with.

The final harmful result for the public when conspiracy theories are used as political weapons is a consequence of the previous ones. The increase in vitriol that we see in our discourse with one another and the general anger that citizens often have for those in different political groups, makes it the case that politicians are more likely to earn votes by satisfying our anger than by actually being good legislators. For example, United States Senator Mitch McConnell is actually celebrated in conservative circles for denying then President Barack Obama the opportunity to bring his choice to replace United States Supreme Court Justice Antonin Scalia up for a Senate confirmation hearing and vote. In previous times this would have hurt McConnell politically, as it would have been viewed as a dereliction of duty. In the current climate it is viewed as McConnell's being a good politician. More than one commentator referred to this as "owning the libs." Given our current tribalism-based political situation, good statesmanship is simply not valued in any real sense.

Fake News and Alternative Facts

As we've been suggesting, our current predicament with respect to trust in our political leaders, political institutions and experts is pretty dire. It's worth exploring how we got here, if only to see what we might do to get out of the predicament.

The distrust and vitriol described above has been present for quite some time (at least as long as the United States has existed) and there is some good reason for this. Politicians have on many occasions engaged in unethical practices and been part of conspiracies. Dead persons have occasionally cast votes, ballot boxes have occasionally been stuffed, bribes have been accepted, opposing campaign headquarters have been bugged. The list goes on and on, but you get the idea. The activities of politicians have made it such that citizens are right to be dubious of both their intentions and their actions. A healthy amount of skepticism is not only warranted, it is needed if we are to keep our democracy on track.

This healthy skepticism does not, however, entail that we ought to believe conspiracy theories. Nor does it, by itself, appear to motivate belief in conspiracy theories. For example, while the assassination of President Lincoln led some folks to believe conspiracy theories pertaining to Lincoln's death, most people did not believe that the Pope, Edwin Stanton, or Jefferson Davis were behind the killing; they accepted the official story, as it were. And yet, nowadays when nearly any political or historical event occurs, we find a sizeable number of people willing to believe that a conspiracy was afoot.

At least part of the reason for this lies in the fact that over the past four-plus years some of our leaders have been working hard to discredit both the press and generally accepted experts. Now if some person believes some proposition, and one can demonstrate to that person that the proposition in question is false by citing a credible source, instead of the usual revising of positions, we frequently find cries of "fake news" or appeals to "alternative facts." As we've seen, this is how conspiracy theories work. Nothing can count as falsifying them, because everything that seemingly tells against the theory is part of the conspiracy. Sadly, this way of thinking has been extended beyond conspiracy theories. A willingness to disregard any authority on any matter, has led us to a point in which political discourse on almost any issue is unlikely to be fruitful. So, the political discourse just gets replaced by more vitriol and grandstanding.

Getting out of this predicament will be no easy task, but it cannot happen until faith in the role of experts (including the press) is restored. Electing honest politicians is probably a good first step.

9
Conspiracy Theories, Social Media, and the Internet

The Flat Earth Conspiracy Theory

Pluto is not particularly interesting. It is mostly ice and rock, it is uninhabited, and it is only about one-third the size of the moon. Perhaps the most interesting thing about Pluto is that people on Earth cannot make up their minds about whether it is a planet. (For what it's worth, we're rooting for you, little guy!)

The Earth, by contrast, is a most interesting place. We've got ice and rocks, but also water, an atmosphere, plants, animals, seasons, ecosystems, fossils, communication systems, rugby teams, Mardi Gras festivals, laptop computers, David Lynch, ukuleles, Stonehenge, Hoagy Carmichael songs, *The Great British Baking Show*, Disneyland, and Yamazaki fifty-five-year-old scotch.

One interesting thing about the Earth is its shape. It is approximately spherical, hence making it a spheroid. This is interesting because from many, if not most, vantage points on Earth, it does not appear this way; rather, it appears flat (or mostly flat plus some hills, mountains, and valleys, etc.). The discrepancy between the way the Earth appears and the way it actually is, has to do with its being very large.

Around 500 B.C.E., Greek philosophers Pythagoras and Anaxagoras theorized that the Earth was "round" (meaning spheroid). Pythagoras "deduced" the roundness of the Earth from observations about the shape of the moon, and Anaxagoras used the shape of the Earth's shadow during lunar eclipses to conclude that the Earth was a sphere. A couple of centuries later, astronomers were able to demonstrate that the Earth was roughly spherical and were even able to calculate the Earth's circumference.

Subsequently, the shape of the Earth has been demonstrated to be spherical literally thousands of times, by (among other things) ships and aircraft that have circumnavigated the globe, satellites, rockets, space shuttles, and space stations that have orbited the Earth, and by photographic evidence provided by said satellites, rockets, space shuttles, and space stations. Prior to the Earth's having been demonstrated to be a sphere, many folks believed that the Earth was flat, sort of like a disk.

While the evidence for the proposition that the Earth is not flat is overwhelming and conclusive, by now it should come as no surprise that there are many who still cling to the belief that the Earth is, in fact, not a sphere, but rather is flat. For some the old adage "Seeing is believing" is wrongly taken to mean that we should always trust our senses. Oddly enough these same folks don't believe other things their senses tell them, such as that the sun and the moon are about the size of a quarter or half-dollar, that the stars are about the size of a speck of dust, and that things are sometimes actually blurry, as when you're not wearing your spectacles or when you first wake up. The "Seeing is believing" set comprises a small subset of those who embrace the Flat Earth hypothesis. Most flat Earthers, instead, subscribe to one of several flat Earth conspiracy theories.

The version of the Flat Earth conspiracy theory that is currently in vogue maintains that the Earth is a disk that has the Arctic Circle at its center and is enveloped by Antarctica. A several hundred-foot-tall wall of ice in Antarctica keeps objects from falling off the edge (picture something like The Ice Wall from *Game of Thrones*, only instead of keeping the riff-raff out it keeps it in). The Sun, on this story, is merely a giant spotlight. Regarding the mountain of evidence to the contrary, Flat Earthers claim that GPS data is manipulated to trick pilots into believing that they have circumnavigated the globe, and that NASA has faked the photographic and other evidence. Each of these claims is easily refuted, but as is always the case with conspiracy theories, the refutations just get woven into the story behind the conspiracy—the science that refutes the theory is false science propagated by those in on the conspiracy.

While there have pretty much always been Flat Earthers, their numbers have been on the rise of late. Currently The Flat Earth Society Facebook group has over 200,000 page likes, and this one is far from the only flat Earth social media group. The rise in the number of flat Earthers can be accounted for primarily by 1. an increase in the number of people who are predisposed to accept any conspiracy theory that asserts that the government has engaged in conspiratorial behavior, and 2. the role of social media in the dissemination of conspiracy theories.

Oh, What an Interweb We Weave . . .

As was mentioned in Chapter 1, conspiracy theories, thanks to the Internet, now have the ability to spread like wildfire. A conspiracy theory can take root and spread around the world in a matter of minutes.

There's an interesting irony at play here. On the one hand, the Internet has rightly been hailed as the greatest source of information ever. You can come to know almost anything you want to know immediately by simply performing a Google search on your given topic.

Obviously, there is a little hyperbole going on here. For example, detectives can't simply enter "who committed the murder I'm currently investigating?" into a search engine and hope to get a correct answer, nor can we glean facts about things that are not already documented, such as "What color socks will George Clooney wear next Thursday?" Still, the Internet does put an extremely large amount of information right at our fingertips.

This ability puts people in a position to either confirm or disconfirm virtually any claim that they are presented with. Moreover, doing so doesn't require a ton of work. There are countless Internet destinations doing the work for us. For example, we could check sites such as Snopes, FactCheck, or PolitiFact to see if some purported event really occurred or if some person made some particular claim. Similarly, we could verify scientific information, such as that the Earth is not flat, with a very quick Google search. Given this, one would expect that on balance the internet would be a force for good in the fight against conspiracy theory promulgation. On the other hand, despite the fact that the Internet is this incredible source of easily obtained true information, it turns out not to be the case that the Internet is a force for good when it comes to stopping the spread of conspiracy theories. This is because the small but growing segment of the Internet that is social media plays an absolutely devastating role with respect to the various efforts to keep conspiracy theories in check.

So how and why does this happen? Why does the Internet end up facilitating the spread of conspiracy theories as opposed to mostly working to stop it (again, given the vast amount of fact checking resources available)? The simple answer to these questions is that there is no simple answer. It is complicated. A number of factors play a role. Moreover, without a number of these factors at play, the Internet would not be a force for evil in the war against conspiracy theories. Suppose, for example, that a random person with a great number of friends or followers on social media decides to start a conspiracy theory. For the purposes of illustration, we call this the "Lizards Aren't Real" conspiracy theory, and stipulate that it mirrors the Birds Aren't Real conspiracy theory in all the relevant aspects (same backstory, same types of evidence, same set of conspirators, and so forth). Further suppose that this random person's followers run the gamut from uneducated to well-educated, and are represented by all parts of the political spectrum. Given these facts, it's unlikely that a handful of posts proclaiming that lizards are not real would lead to any widespread acceptance of the conspiracy theory. Rather, most of the person's friends and followers would ignore it, while some would take issue with the claims. Possibly a few would accept it. The result is that someone merely asserting a conspiracy theory on social media is not by itself sufficient for it to go viral, even if it is asserted to a large number of people. Some or all of the aforementioned, but not yet explicated, factors need to be in place, in order for a conspiracy theory to take off on social media. So, what are the factors?

The first factor is that the conspiracy theory must be well-formulated. As we've seen, a successful conspiracy theory must tell a plausible story about how the conspiracy can be true, and how the prevailing narrative is false. The bar for what counts as "plausible" is pretty low. Conspiracy theories can involve appeals to supernatural creatures and can cast entire categories of seemingly perfectly normal people into a horrid light. So, what we mean by plausibility here is that the theory must count as something that in principle could explain the facts, even if that explanation is far-fetched. Since we've covered this quite a bit already, and this factor is not specific to social media, there is no need to say more about this one here.

A second factor lies in the fact that increasingly people are turning to social media as their primary news source. In fact, the number of people in the United States who get their news from social media has doubled in the last eight years (Stekula and Pickup 2021). One problem is that the standards for news sources are quite variable. While one might access credible news sources on social media that employ the highest degree of journalist standards (think the good old "Fourth Estate" here), one might just as easily access news sources that employ virtually no standards whatsoever (other than those non-journalistic standards that serve to ensure influence or revenue). This means that people will be subject to misinformation that is presented as vetted fact. Moreover, this information in terms of presentation might appear in a similar fashion to traditional news, for example, with headlines or a byline, similar wording ("News Flash: Scientists Report . . ."), and similar graphics and illustrations, thus making it even harder to distinguish from credible information. A second and related problem is that most people are just not trained to distinguish good information from bad, nor are they trained to distinguish credible sources from non-credible sources. Of course, others are simply not willing to put in the effort to distinguish the credible from the non-credible. As President John F. Kennedy famously observed "Too often we enjoy the comfort of opinion without the discomfort of thought."

People's inability or unwillingness to do the work necessary to distinguish credible sources from non-credible sources has always been a problem, but it's further compounded by the fact that certain politicians have taken to discrediting the credible sources whenever those sources report things that are not to the politician's advantage. Cries of "fake news" have created a real crisis among those who actually do want to distinguish credible news from non-credible news, but lack the tools to do so.

A second factor lies in the fact that news on social media tends to be curated, meaning that people get news that is selected for them based on things that they are already known to believe. This entails that people are more likely to get news that they either already believe or are more likely to believe—they get the news that they want! Recall that in our hypothetical scenario, the person promoting the Lizards Aren't Real conspiracy was delivering claims to a wide variety of persons (educated, uneducated, liberal, conservative, and so on). Those who promote conspiracy theories in the real world have the ability to target their information, which again, comes disguised as credible news, to precisely those who are more likely to believe their conspiracy theories.

Selective targeting happens in at least two ways. First, since people tend to associate with like-minded people, the information that gets shared tends to get shared with people who are similarly disposed to accept the information, with only small bits of information being shared with those who might be inclined to either reject it, or worse from the perspective of the sharer, argue against it. Second, it is not just people who are sharing the information. Bots (internet robots) are employed to disseminate information (or misinformation or fake news or conspiracy theories) to social media users by posing as real people, then rapidly sharing the information that they are designed to spread. Moreover, algorithms utilizing data gathered from other bots along with other data collection tools identify precisely who the bots should attempt to friend or follow, etc. on social media, to ensure that the information reaches precisely those who are most apt to believe it.

If you're reading this and thinking to yourself, Gosh, I sure get a lot of conspiratorial information, it almost certainly means that the bots have your number. You may want to reconsider several of your positions, as well as take a look at some different news sources (you may also want to take a close look at your friend circles, but we recognize that such things can be complicated). On the other hand, if you don't see much of this sort of thing on your social media feeds, then it's likely that you have been identified as someone who is not likely to accept conspiracy theories. You're doing well in this regard. Keep up the good work! (A third option is that you don't recognize conspiracy theories when you see them, but we'll set that option to the side).

A third factor has to do with whether we generally trust what we find on social media. A couple of recent studies show a positive correlation between trust in social media and likelihood of accepting conspiracy theories (McGreal 2021; AAAS 2021). Given this fact, as people increasingly trust and get their news from social media, we would expect to see an increase in the rate that conspiracy theories spread and gain acceptance. This isn't the case for all social media platforms. People who get their news from WhatsApp, YouTube, and Facebook, according to the data, are more likely to accept conspiracy theories, but those who primarily get their news from Twitter actually become less likely to believe conspiracy theories. Researchers speculate that the particulars of Twitter's forum make it such that users will in most cases be exposed to a wider variety of viewpoints and news sources. This serves to mitigate the effects of bots.

The final factor in social media's ability to rapidly spread conspiracy theories is confirmation bias. As we've seen in previous chapters, people are greatly subject to confirmation bias. If the bots are doing their jobs right, and spreading information to people who already believe similar things to what they currently want them to believe, then confirmation bias will serve to virtually ensure that the targeted people will believe what the bots are designed to get them to believe. In short, people are being targeted with information that they already strongly desire to believe.

Still More Badness

We've already discussed at great length why acceptance of conspiracy theories can be a bad thing, so for the most part, the spreading of conspiracy theories on social media is only bad to the extent that people accepting the conspiracy theories is bad in and of itself (regardless of how people came to accept some particular conspiracy theory or other). Moreover, in instances where accepting the conspiracy theory isn't all that bad (for example in the cases of the Birds Aren't Real conspiracy theories), the fact that social media played a role in spreading the conspiracy theory doesn't serve to make the conspiracy theories any worse. That said, there are instances in which the rapid spread of conspiracy theories due to the role that social media played, certainly exacerbates the harms.

A case in point is the QAnon endorsed sex trafficking conspiracy theory that populated the Internet via memes and hashtags (#savethechildren) in the late summer and fall of 2020. This was a coordinated effort to 1. distract from other issues at the time, perhaps most importantly then President Donald Trump's failure to deal with the coronavirus pandemic in any appreciable way, and 2. politicize the sex trafficking of children—this particular campaign attempted to make it appear as though Democrats had no interest in addressing the issue. It's no coincidence that these memes were coming out during the final months of the 2020 Presidential election. This was supposed to function as a sort of "October surprise."

So far, this all sounds good. What harm could possibly come from shining a light on the trafficking of children for sexual purposes? Unfortunately, the consequences of this campaign were quite bad in two respects. It actually made it significantly harder for crime fighters to do their jobs. Angry QAnon supporters literally flooded law enforcement tip lines with false leads, increasing the amount of work that needed to be done. Moreover, the average quality of the leads investigators were provided with during this period of time were considerably worse. The good leads got buried in the process, and the resources of various law enforcement agencies were rapidly drained. The second bad consequence was an increase in vigilantism, due to the messaging that the problem was being ignored. This by the way is a great example of the dangers of the fallacy known as "Complex Question." The posts and memes asked the question "Why is no one doing anything about child sex trafficking?" Of course, much was being done at the time, but it led people to believe that nothing was being done. All said and done, had this conspiracy theory not spread the way that it did across social media, none of these bad consequences would have resulted.

What Can Be Done About It?

It is clear that social media constitutes a powerful tool for those who desire to promote conspiracy theories. This raises the question: what can be done about it. The situation is pretty dire, at least in the short term, but if the proper steps are taken, over time, there is reason to be sanguine about mitigating the role that social media plays in spreading conspiracy theories.

Here are the steps.

First, we need to educate people about conspiracy theories. People need to know what they are, how they work, and why they are harmful. Many people (perhaps most) are under the impression that any theory about a conspiracy is a conspiracy theory, and since conspiracies happen with some regularity, many feel that they have a reason to believe any conspiracy theory of which they are made aware. As we've seen, this couldn't be farther from the truth. Belief in true, bona fide conspiracy theories, that is belief in conspiracy theories as we've defined them in this book, as opposed to mere theories about conspiracies, is almost never warranted.

Second, we need to educate people about how social media works. People need to know that not everything that looks like a news source, is, in fact, a credible news source. People also need to be made aware of the ways in which the information they receive is curated precisely for them, and delivered with pinpoint accuracy by bots. Relatedly, people need to be made aware of their cognitive biases. An awareness that one is predisposed to believe certain things is a necessary step in one's becoming guarded against accepting such things (at least without doing one's epistemic work first).

Third, we need to put more pressure on social media providers, such as YouTube, Facebook, WhatsApp, and Twitter to stop the spread of conspiracy theories. Over the past twelve months we've seen a number of policy changes in this direction, but more needs to be done. These providers need to be even more aggressive in slowing the spread of conspiracy theories.

Finally, we need to better educate people about how to distinguish good information from bad. A first step in this direction might involve teaching people about how to verify claims they come across on the Internet (for example, how to use resources such as Snopes, FactCheck, and PolitiFact). One worry here is that each of these sources has been the subject of conspiracy theories. Anecdotally, we can report that we've witnessed a number of people claim in response to having their facts checked, that the fact checking sources themselves are biased and therefore not credible. Those who hope to slow the spread of conspiracy theories over social media are certainly facing an uphill battle.

10
Existential Matters and Categories of Conspiracy

The Death of Marilyn Monroe

On the morning of August 5th 1962, Hollywood actress Marilyn Monroe was found dead in her bed. She was nude with a telephone in her hand. Bottles of pills, which she had been prescribed for depression, were found scattered in her bed and around the room. She was known to regularly consume drugs and alcohol, often at the same time. Her death was ruled a suicide. Her body was subsequently cremated.

Shortly thereafter, people began to express disbelief about the whole set of circumstances. Some maintained that there were reasons to believe that the death of America's sweetheart was not self-inflicted but was instead a murder. Others insisted that she was not dead at all—the whole thing was a ruse to provide the actress with a path out of a miserable life and an escape into anonymity. Friends reported that prior to her death, she seemed to be happy and had made plans with loved ones for activities in the future. That said, Monroe was known to suffer from depression. One thing all of the conspiracy theories shared in common was a reluctance to believe that someone as beloved, beautiful, and successful by all metrics deemed important by society at the time as Marilyn Monroe really could have killed herself.

Those who believe that Monroe was murdered do not all agree when it comes to the perpetrator or the motive. Some have argued that the murder was related to her relationship with the Kennedy family. Rumors have long circulated that Monroe had a relationship with John F. Kennedy, Robert Kennedy, or both. Regardless of who the alleged affair was with, the story is that when she seemed too attached, the Kennedys had her killed to protect their political careers. In the second installment of *American Horror Story*'s two-part tenth season, the writers float the existing conspiracy theory that Monroe was killed because JFK told her too much about encounters with aliens at Area 51.

Others believe that Monroe's death was a Mafia hit. One theory of this type is that Jimmy Hoffa ordered the killing in order to hurt the Kennedys. At the time, they were being investigated as part of a crackdown on organized crime. Some of the pills that were found in a bottle next to her bed were chloral hydrate—a drug that was allegedly used in other Mafia hits. Some also speculate that the Mafia hit was related to Monroe's relationship with Frank Sinatra.

Conspiracy Theories and Death

In this chapter, we'll provide a taxonomy of the general categories of conspiracy theories. The topics of conspiracy theories are as broad as the range of events that human beings can contemplate, but, unsurprisingly, such theories tend to coalesce around a handful of key themes. We'll argue that such themes are as much or more instructive about human psychology as they are about questionable alternative explanations for world events. People may cluster around certain themes when they construct conspiracy theories for similar reasons that they are inclined to commit certain kinds of falla-cies—there are some assumptions about life and the way we want it to work that are too valuable to the safekeeping of our psychologies for us to easily or will-ingly give them up. As we will see, these categories do not divide cleanly. Many theories have elements of more than one of these categories. This is to be expected, since the fundamental meaningful aspects of human lives tend to overlap.

Monroe's is far from the only celebrity death that people have refused to accept for one reason or another. Elvis Presley, the "King of Rock'n'Roll" died on August 16th 1977. He died of a heart attack on the floor of his bathroom. Experts believe that the event was precipi-tated by the performer's frequent use of barbiturates. People refused to believe that a figure as prominent as Presley was dead, and they certainly struggled to be-lieve that such a person's death would be so inauspi-cious. James Dean died in a car accident in 1955; plenty of people believe that he didn't die at that time at all, but instead faked his own death to escape the spotlight. Some also believe that Michael Jackson's death due to cardiac arrest in 2009 was faked. People had the same reaction to the deaths of Jim Morrison, Kurt Cobain, Princess Diana, and John Lennon.

Nor is this anything approaching a modern phenom-enon. Historically, the fact that a person had died might be challenging to learn, depending on one's circum-stances. It might be even more challenging to confirm. One historical example of this concerned the Roman Em-peror and philosopher Marcus Aurelius. When false re-ports circulated that Aurelius had died, Avidius Cassius started a campaign for emperor which turned into a po-tential usurpation when it was revealed that Aurelius was still very much alive. In an age without live media coverage or the ability to move quickly from one place to the next, these things were much more difficult to verify.

This created a breeding ground for conspiracy theories that plagued historical figures in ways that resemble the ways that such theories affect celebrities and world leaders today. Consider just one set of events that transpired in medieval England during the Wars of the Roses. The marriage of Henry VII to Elizabeth of York was intended to unite the warring houses of Lancaster and York and to bring an end to the conflict for the throne. Unfortunately, there was the pesky issue of the "Princes in the Tower"—the two sons and heirs of Edward IV. Shakespeare alleges that Edward's own brother, Richard III, ordered the murders and that the bodies were buried on the grounds of the Tower of London. Of course, there was no way to confirm their deaths, so Henry VII's reign was haunted by conspiracy theories contending that one or the other of the princes was still alive. If this were the case, the living son of Edward IV would be the rightful heir to the throne. One noteworthy pretender, Perkin Warbeck, claimed to be Prince Richard and led an uprising against Henry VII. He later admitted to being a Flemish imposter and was executed for treason in 1499. His absurd claim was entertained for so long because of the penchant people have to believe conspiracy theories, especially about death, and particularly when a death conspiracy turns out to be in the interests of the believer for one reason or another.

Death makes people uncomfortable, to put it mildly. The human tendency to think of illness and death as misfortunes that happen only to other people is a form of bad faith that is discussed at length in existential literature. The idea that death will happen to those that we love and admire or, heaven forbid, to us personally, is something that we like to keep at the level of abstraction.

Tolstoy explores these themes in *The Death of Ivan Ilyich*. The titular character suffers a minor accident which leads to his unexpected and untimely demise. He discovers with horror that he must die alone—no one around him is having his set of experiences, so no one can empathize with what he is going through. His friends and family can't relate because they live their lives in denial of the possibility of their own respective deaths. Tolstoy describes the reaction of one of Ilych's friends and colleagues, Peter Ivanovich, on the occasion of his funeral,

> "Three days of frightful suffering and then death! Why, that might suddenly, at any time, happen to me," he thought, and for a moment he felt terrified. But—he did not himself know how—the customary reflection at once occurred to him that this had happened to Ivan Ilyich and not to him, and that it should not and could not happen to him, and that to think it could would be yielding to depression which he ought not to do . . . After which reflection Peter Ivanovich felt reassured, and began to ask with interest about the details of Ivan Ilych's death, as though death was an accident natural to Ivan Ilyich but certainly not to himself.

There is something relatable about Ivanovich's response to his friend's death, but when Tolstoy presents it to us in the third person we can't help but to recognize the absurdity of it. Of course death will come for Ivanovich—as it will for us all. Denial doesn't change anything. Yet, denial is a common response when death and devastation surround us. Death comes for Ivan Ilych, as it came also for Marilyn Monroe, Elvis Presley, James Dean, Jim Morrison, Michael Jackson, and the Princes in the Tower.

Our tendency to believe that celebrities are immune from death is connected to our tendency to commit the *denial* fallacy—the error in reasoning that causes a person to conclude that the fact that they don't want something to be true is sufficient to make that thing false. Even today, people become obsessed with long dead celebrities and refuse to believe that they're dead, or, at the very least, that the way that the icon in question died could have been as mundane or even humiliating as the cause reported.

If the King of Rock'n'Roll could die of a heart attack on the floor of his bathroom, how might a star that burned less bright die? How might *we* die? For many, the question is psychologically too difficult to answer, so they reject the premise.

Consider the public response to the death of Amelia Earhart. Earheart was a national hero and was an incarnation of goals for liberation and career success for women across the world. Eighteen months after her plane crashed on her journey across the Pacific, she was declared legally dead. Many did not want to believe that someone like Earhart could die in that way and concluded instead that she was shipwrecked, eventually saved, and lived on to serve as a spy in Japan during World War II for President Roosevelt.

Conspiracy Theories and Major Tragedies

Conspiracies surrounding major tragedies unsurprisingly share features in common with conspiracies about the deaths of individual people. The expression "major tragedies" could refer to many different kinds of events. In this case, we're using it to refer to injury or loss of life on a major scale. So, though it may have been a major tragedy when David Bowie died, that's not the kind of major tragedy that we have in mind here. We've discussed some of these tragedies already and will do so again in subsequent chapters, but here we'll discuss some of the potential existential motivations for such conspiracies.

Natural disasters are often the source of conspiracy theories. Since such events pose major challenges to life and well-being; scientists have made them the object of study for many years. They have developed ways of determining when disasters of certain kinds will take place. At the recommendations of scientists, local or national authorities can then advise citizens to prepare or to evacuate to minimize the harm done by the disaster. These predictions can be made about earthquakes, hurricanes, tornadoes, and other major earthly events at varying degrees of proximity to the event, depending on the kind of natural phenomenon in question.

Perhaps because the method that leads to the predictive power of scientists is either not transparent to everyone or not investigated by everyone, some people use predictions as evidence of the existence of a plot. Many conspiracy theorists believe that the government has ways of controlling the weather. According to these people, extreme weather events happen as a way of disempowering and manipulating citizens who have become targets for the government.

Again, the motivations for belief in this kind of theory are existential. It's hard to make sense of the suffering that occurs in the world, especially when it seems impossible that the suffering in question is necessary for some greater good. Natural disasters are perhaps the most extreme case of this because they do not come about as a result of the actions of free creatures who exercise their wills to do evil and bring about chaos and destruction as a consequence. Instead, natural disasters, like the senseless deaths of those that we respect and care about, remind us of the absurdity of the human condition.

In *The Myth of Sisyphus,* Camus defines absurdity as a confrontation between the individual and the indifferent universe. Individuals have needs and desires, hopes and plans. The universe isn't the kind of thing that is capable of caring what individuals want. Natural disasters are paradigm examples of the indifference of the universe—they don't avoid destroying houses or beloved community buildings. They don't care if and who they kill. They are just some of the workings of nature, behaving as if sentient beings do not exist. Conspiracy theories about such events are a way of giving individuals someone to blame. For some, it is easier to come to terms with the bad intentions of nefarious agents than it is with the fact that we live in an uncaring universe.

Conspiracy Theories and Technology

Philosophers, for better or for worse, have frequently pointed to the human capacity for reason as the characteristic that sets humans apart from non-human animals. Human beings don't just take the world as they find it, they have the problem-solving skills to adapt nature to fit their own purposes. In many ways, technology is about power. Early in human history, we were powerless against the awe-inspiring forces of nature. Though threats still remain, in affluent areas of the world, humans can largely now protect themselves against nature's most significant threats. As a result of technology, and the industrial revolution in particular, humans have, in many ways, turned the tables—we have become a threat to the stability and security of the natural world.

Technology creates a noteworthy power dynamic between humans and nature, but it also magnifies differences in access to power between individuals and groups. Among other things, rich and powerful people have had access to technology that poor people have not—the rich have had access to different medical treatments, different, faster and more convenient forms of transportation, better forms of technology to promote and spread messages, and different mechanisms to grow their wealth and power. It's no wonder, then, that what people often perceive to be the source of power imbalance—emerging technology—has also been the source of tremendous incredulity by those on the losing side of that dynamic.

Our abuse of the climate through greenhouse gas emissions is beginning to cause the harm that has long been predicted. In response to this, scientists are doing their best to make the situation better. Though it may now be impossible to completely undo what we've done, there are steps that we can take to make the consequences less bad. These approaches uniformly involve emerging technologies and these technologies inevitably lead to conspiracy theories. We'll describe some of these technologies below and then look at the conspiracy theories that have emerged in response.

Scientists have long been developing methods of green energy production, though political gameplaying has prevented these technologies from being adopted as the dominant forms of energy. Industrial animal agriculture is a significant cause of climate change. The animals themselves in great numbers produce significant amounts of methane and other greenhouse gases. The processing of animals does the same thing. On top of all of this, deforestation to make room both for animals to graze and for grains to feed animals eliminates trees—critical for cleaning the carbon from the air. The bottom line is—we need to abandon or, at the very least, substantially reduce our reliance on industrial animal agriculture.

Food scientists have been hard at work creating alternatives. Singapore has recently become the first country to give regulatory approval to the sale of *in vitro* meat (BBC News 2020). This process generates meat without killing any animals. Instead, cells are harvested by way of a biopsy. The cells are then taken to a lab where they are fed with growth serum and placed in a bioreactor. The result is that scientists can create meat of essentially any type (though the architecture of some forms of meat poses some practical problems). Here we have a way of feeding a planet that seems insistent on continuing to eat meat while at the same time raising significantly fewer animals (and none of them for slaughter!).

Scientists also respond to food scarcity and growth challenges by genetically modifying some plants and other organisms. Genetic modification of plants has different purposes depending on the crop. In some cases, it can help keep fruits and vegetables from bruising, it can make the plant immune to herbicides so that weeds can be controlled without hurting the crop, it can protect against plant viruses, and it can, in some cases, protect against the need for insecticides. Genetic modification has the potential to help farmers to grow plants in greater abundance. This is helpful for a world in which food shortages are extremely common in certain regions.

Meanwhile, some scientists and philosophers have grown quite pessimistic about the likelihood that human beings will or even can take the necessary steps to roll back climate change in time to prevent all sorts of cataclysmic events—lives lost, climate change refugees struggling to find a safe place to live, and countless species of animals, essential for the well-being of ecosystems, gone forever. In response, they develop and argue for various methods of geo-engineering. One form of geo-engineering involves what is commonly referred to as "solar radiation management." This approach involves technological strategies designed to control solar radiation. One such proposal involves mirrors strategically positioned in space that are designed to deflect solar radiation back out into space. Other such strategies involve injecting particles into the atmosphere that make it more difficult for radiation from the sun to penetrate to the areas below. Other strategies involve carbon dioxide removal. All of these projects involve advanced and expansive technology.

If the technologies described above are implemented, some aspects of the world as we know it will change quite a bit. Other aspects have a much better chance of staying the same than they would if we did nothing.

As you might have anticipated, conspiracy theorists have gone to town with all of this. Many YouTube personalities are describing these climate change related technologies, taken together with the science that has been developed to combat the pandemic, as "The Great Reset." The idea is that both the pandemic and climate change are hoaxes that have been created in order for the rich and powerful (those are often euphemistic terms used to describe Jewish persons by such people) to exercise complete control over the entire global population. Here's Bill Gates again, or maybe more often George Soros, trying to completely control food, or weather, or climate, or . . . really anything that can conceivably be controlled.

The idea that government could try to control its citizens through the use of technology is not new. It is a common theme in dystopian literature and other forms of dystopian entertainment: the life-size television characters in Bradbury's *Fahrenheit 451*, the biotechnology in Atwood's *Oryx and Crake*, and Bokanovsky's Process, birthing pods, and soma in Huxley's *Brave New World*. Most episodes of Netflix's *Black Mirror* involve some form of government manipulation using technology. People have ready and constant access to fictional tales that portray governments seriously misusing all sorts of technology that is either similar to or identical with the technology that we've developed in the real world. What's more, the protagonists that challenge the status quo in these stories are doing what the reader or viewer is encouraged to think of as the "right thing." With all of this motivation and inspiration, it's not surprising that technology skeptics think they're onto something in the real world.

Manipulation through the use of technology is clearly not merely the stuff of fiction either. All readers will likely be familiar with the phenomenon of shopping online for an item only to have items of that type advertised, unsolicited, on social media. Companies like Facebook, Amazon, and Google, use our private data in all sorts of ways on a regular basis, frequently with something adjacent to "consent" from us—we'll willingly click "accept" when Google asks us if they can track our location without wondering if that means that they'll share that data with many of those who are willing to make it worth their while.

It isn't just corporations or nefarious tech-savvy individuals engaged in the harm or manipulation, either. Governments have a track record of using technology to bad effect—sometimes deceptively and sometimes right out in the open. In the 1960s, environmental writer Rachel Carson brought to a stunned and incredulous public audience's attention the fact that the government was regularly using dichloro-diphenyl-trichloroethane (DDT) as a pesticide. DDT was developed partially during the second world war and its potential for insect control was enthusiastically pursued by the US. It quickly became clear that the pesticide was a poison for more than just insects. It was killing animals, ecosystems, and even human beings at alarming rates.

Some government abuse of technology has been targeted at particular races. Consider a famous case in the history of medical experimentation—the Tuskegee experiments. The government had treatments for syphilis, but they wanted to monitor what would happen if the disease was allowed to progress completely untreated. To do this, they told a group of black men in Tuskegee that they were in a trial group for a new medicine. The pills they were being given were actually placebos. The men then suffered the effects of untreated syphilis until many of them died. In this case, the citizens were not directly harmed by technology, but were instead harmed and manipulated by deception related to the withholding of the technology.

There are many more abuses or potential abuses of technology by the government—so many that it would be impossible to provide anything close to a full accounting of them here. Infrared and drone technology have changed the nature of war and threaten the protections offered by the fourth amendment. Governments use online manipulation technology to affect the outcome of the elections of rival or enemy countries. All this is to say: concerns about private and governmental overreach when it comes to technology are far from unwarranted.

Our uneasy relationship with technology, the power dynamics that it is capable of exploiting to great effect, makes new frontiers and developments in technology ripe for conspiracy theorizing. This concern isn't merely contingent; it's existential and arises out of the fact that one of the essential human survival mechanisms is to solve the problems that we encounter with tools of ever-increasing complexity and capacity for violence.

The problem is that not all technology is the proper source of concern. Much of the technology that is currently being developed, including the technology that believers of "The Great Reset" conspiracy theory are so concerned about, is technology that is being developed to save humanity from the biggest challenges that it faces, not to control people.

One diagnosis of the problem here has to do, once again, with trust in expertise. The average person is not familiar with the lifestyles and motivations of academics. Many are not familiar with peer review or academic journals. They assume that the motivations of small, local players who work at universities or in labs are the same motivations that they take Soros and Gates to have—to obtain and wield unprecedented amounts of power and control. Instead, the motivations are often to do their part to bring about good in the world, get tenure and promotion, and perhaps make a modest amount of money along the way (depending upon the technology in question, of course, and the way in which it is researched). Nevertheless, because the average person is removed from the process of developing new technology, they may be in a diminished position to determine how and why it is being developed. Technology is essential for continued human life on this planet, but advanced technology means increased power, and that scares people on a fundamental level. For this reason, technology-related conspiracy theories, though they overlap with others, deserve a designation as a category of conspiracy theory on their own.

Conspiracy Theories, Subordination, and Power

Emerging technology is one aspect of life that can draw attention to imbalances of power. There are many others: socio-economic imbalances, epistemic injustice (a cluster of phenomena which involve, among other things, some voices being listened to more than others), imbalances of goods and services, and so on. Understandably, we don't like to end up on the losing side of these dynamics, and not only because we want to avoid suffering. Some philosophers, like Friedrich Nietzsche, have argued that human beings are driven by a will to power, though exactly what that means is open to debate. People long to be taken seriously, to feel like their interests and perspectives matter, and to feel that they are each an individual with dignity rather than simply a cog in a meaningless social machine. Many feel a further need to dominate and to "win" the games of life—to maximize their own acquisition of the earth's resources, prestige, and other advantages. If there really is such a will to power, some might satiate it with self-mastery, the careful development of virtuous character traits, and the power to do what one can to help others. The point is—we strive to avoid impotence and doing so matters to us on an existential level.

However one conceives of power, people often do not do well when they lose control. When people are put in situations in which they find their power diminished, when they feel minimized or ineffectual, a conspiracy theory can help them to deal with their feelings of inferiority and insecurity. Losing the game because someone cheated can be easier on the ego than losing a game for lack of skill or because hey, sometimes you just lose.

This fundamental human desire to understand oneself as, at least in some sense, powerful might play some part in explaining why people are quick to blame immigrants for job loss or racial minorities for changes in social policy and one's sense of social stability.

One noteworthy example of a conspiracy theory of this type is the theory frequently referred to as "The Great Replacement." The idea is that one racial or ethnic demographic in a particular country is *intentionally* being replaced with another racial or ethnic demographic in order to change the nature of social policy and the political balance. This view was popular in the twentieth century in multiple countries across Europe and it remains quite popular in the United States and elsewhere. The view is commonly held among members of white supremacist groups and it is the underlying motivation behind the white supremacist mantra "The Fourteen Words" which are the following, "We must secure the existence of our people and a future for white children." Fox News commentator and propogandist Tucker Carlson has recently unapologetically advanced the theory, arguing that democrats are systematically encouraging immigration in such a way that voters of color will outnumber white voters to ensure that Republicans never win another election. The theory often seems to take it as given that immigrant populations will vote for liberal policies.

This conspiracy theory, in addition to providing much of the motivation for the Holocaust, has also been offered as the explanatory reason for much of the domestic terrorism that has taken place in the United States in recent years (ADL 2021). Robert Bowers wrote an online post referring to The Great Replacement as his reason for killing eleven Jewish people at the Tree of Life Synagogue. Patrick Crusius killed twenty-three people at a WalMart in El Paso with an AK-47-style rifle citing concerns about a "Hispanic invasion." John Earnest claimed that the reason he killed one person and injured another three at a California synagogue was his belief that Jews are responsible for non-white immigration into the United States. In New Zealand, killer Brenton Tarrant cited "The Great Replacement" as a rationale for killing fifty-one people at two different mosques.

One explanation for all of this is impotence, a sense of impotence, or both. When a person feels that they are losing the power they once had, they want someone to blame. Racist and xenophobic conspiracy theories frequently fit the bill. As much as it may revolt us, we are again dealing with an issue with existential import. Whether a person is powerful or not is crucially important to that person— it is frequently a status that they would rather die than to give up. It's no surprise, then, that such a phenomenon frequently gives rise to conspiracy theories.

Conspiracy Theories and Identity

Deep differences of opinion about the pandemic, race, and government have created chasms of frustration, distrust, and misunderstanding. If this is an accurate description of relationships between those who cared deeply for one another, it's even less likely to be resolvable for casual acquaintances and members of our communities we only come to know as a result of our attempts to create social policy. The social situation has served to amplify our already significant sense of grief, loss, and loneliness—the comfort of community is gone. We feel what's missing acutely. How ought we to deal with these differences? Can we deal with them without incurring significant changes to our identities? Perhaps the most compelling existential challenge, and one that frequently gives rise to beliefs in general and conspiracy theories in particular, is our understanding of our own respective identities—who we take ourselves to be at our very cores.

Moral philosophy throughout the course of human history has consistently advised us to love our neighbors. Utilitarianism tells us to treat both the suffering and the happiness of others impartially—to recognize that each sentient being's suffering and happiness deserves to be taken seriously. Deontology advises us to recognize the inherent worth and dignity of other people. Care ethics teaches us that our moral obligations to others are grounded in care and in the care relationships into which we enter with them. Enlightenment moral philosophers like Adam Smith have argued that our moral judgments are grounded in sympathy and empathy toward others. We are capable of imaginatively projecting ourselves into the lives and experiences of other beings, and that provides the grounding for our sense of concern for them.

Moral philosophers have made fellow-feeling a key component in their discussions of how to live our moral lives, yet we struggle (and have always struggled) to actually empathize with fellow creatures. At least one challenge is that there can be no imaginative projection into someone else's experiences and worldview if doing so is in conflict with everything a person cares about and with the most fundamental things with which they identify. In our current climate, many people look at believers in the QAnon conspiracy theory as if they have lost their minds and adherents of The Big Lie as if they are the authors of doom for our democracy. On the other side, conspiracists would rather die than become "sheeple." They feel that they have undergone a great "Awakening" and they'd rather die than to come back or to believe as non-conspiracists believe.

Fundamental commitments make us who we are and make life worth living. In fact, the fragility of those commitments, and thus the fragility of our very identities, causes some philosophers to argue that immortality would be undesirable (bear with us, we promise this is relevant). In Bernard Williams's now famous paper "The Makropulos Case: Reflections on the Tedium of Immortality," he describes a scene from *The Makropulos Affair*, an opera by Czech composer Leoš Janá ek. The main character, Elina, is given the opportunity to live forever—she just needs to keep taking a potion to extend her life. After many, many years of living, she decides to stop taking the potion, even though she knows that if she does so she will cease to exist. Williams argues that anyone who takes such a potion—anyone who chooses to extend their life indefinitely—would either inevitably become bored or would change so much that they lose their identity—they would, though they continued to live, cease to be who they once were.

One of the linchpins of Williams's view is that, if a person puts themselves in countless different circumstances, they will take on beliefs, desires, preferences, and characteristics that are so unlike the "self" that started out on the path that they would become someone they no longer recognize. One doesn't need to be offered a vial of magical elixir to take on the potential for radical change—one has simply to take a chance on opening oneself up to new ideas and possibilities. To do so, however, is to risk becoming unmoored from one's own identity—to become someone that an earlier version of you wouldn't recognize. While it may frustrate us when our friends and loved ones are not willing to entertain the evidence that we think should change their minds, perhaps this shouldn't come as a surprise—we sometimes see change as an existential threat.

Consider the case of a person who takes being patriotic as a fundamental part of their identity. They view people who go into professions that they deem as protective of the country—police officers and military members—to be heroes. If they belong to a family which has long held the same values, they may have been habituated to have these beliefs from an early age. Many of their family members may be members of such professions. If this person were asked to entertain the idea that racism is endemic in the police force, even in the face of significant evidence, they may be unwilling and actually incapable of doing so. Merely considering such evidence might be thought of, consciously or not, as a threat to their very identity.

The challenge that we face here is more significant than might be suggested by the word "bias." Many of these beliefs are reflective of people's categorical commitments and they'd rather die than give them up. None of this is to say that significant changes to fundamental beliefs are impossible—such occurrences are often what philosophers call transformative experiences. That language is telling. When we are able to entertain new beliefs and attitudes, we express a willingness to become new people. This is a rare enough experience to count as a major plot point in a person's life.

This leaves us with room for hope, but not, perhaps, for optimism. Events of recent years have laid bare the fundamental, identity-marking commitments of friends, family, and members of our community. Reconciling these disparate commitments, beliefs, and worldviews will require nothing less than transformation.

Sometimes transformation is a good thing and sometimes it isn't. If you're a narrow-minded person who's never much interested in hearing what other people think or considering other points of view, it would probably be a good thing for you to become a little more open minded. If our uncle in a tin-foil hat really wants us to consider the antisemitic idea that a cabal of ill-intentioned Jewish persons are trying to perpetuate a genocide against the white race, we're just not willing to try on the racism. Doing so is antithetical to the core components of our identities. We would rather die than be antisemitic. We want neither racism nor antisemitism to be part of our own life narrative, to the extent that we can possibly prevent it (at least in regard to our own worldviews).

The problem is that conspiracists view such a transformation as similarly unfavorable and identity-destroying. The desperate attempt to defend a conspiracy theory may often be a desperate attempt to preserve our identity and our dignity—the desire to not become a fool through the gaze of the other. So, here we are and here we have always been, to some degree or other.

In the next chapter, we'll dive deeper into a diagnosis of this general phenomenon—we'll turn to the psychology of conspiracy theories.

11
The Psychology of Conspiracy Theories

The Satanic Panic

By the early 1980s, the country had undergone massive changes with respect to the workforce. World War II initiated many women into work outside of the home for the first time. The feminist movement in the intervening years brought increased rights and opportunities for women, and by the 1980s, women entered jobs for which they had carefully planned—they designed the structure of their educations to prepare for them. More women than ever entered the workforce with the expectation that they would spend much of their lives there. Under these conditions, access to quality affordable childcare became a necessity.

A brave new world with women in the workforce was the kind of future that many people found distasteful. The idea that a child would be cared for during the day by people other than that child's mother struck many conservatives as unnatural and contrary to God's wishes. Unsurprisingly, daycare centers were soon the subject of attack for various reasons.

In 1983, a woman accused an employee of molesting her son at his California daycare. In response, the local police asked for the help of parents in the community.

They asked two hundred sets of parents for their assistance with interrogating their children. None of the parents had any experience interrogating a young person about potential assault. In the modern world, we tend to think of that as a task for which a person should be carefully trained and well-educated. Soon, very strange stories began to emerge and the phenomenon called the "Satanic Panic" began, not only across the state of California, but across the whole country. According to the New York Times, allegations from students and parents soon came to include, "a 'goatman', bloody animal sacrifices, a school employee who could fly and acts of violence that left no physical trace" (Yuhas 2021). Professional law enforcement officials held conferences on Satanist symbols and imagery so that colleagues and parents would be in a position to recognize the calling cards. The *New York Times* also reported that, by 1985, allegations of this type were so prolific that 20/20 did a segment describing reports of "animal mutilations 'clearly used in some kind of bizarre ritual', rock music 'associated with devil worship', 'satanic graffiti', and backward messages in pop songs." Before long, there were allegations about Satanic rituals involving child molestation across the country.

Other types of cases entirely were reinterpreted by some investigators through the lens of occult theories. For example, journalist Maury Terry wrote a book claiming that serial killer David Berkowitz, who paralyzed New York City with fear in 1976 and 1977, did not work alone as The Son of Sam, but, instead, was part of an elaborate satanic conspiracy to kill women for ritualistic purposes (Terry 1988). See what happens when women enter the workforce instead of staying home and taking care of their children? Their kids get abused by Satanists and the whole country falls victim to occult forces.

Except, of course, none of this was true. Despite the alarming number of allegations, there was no evidence to substantiate any of it. In 1994, researchers from the National Center on Child Abuse and Neglect studied twelve thousand allegations of satanic ritualistic child sex abuse and found no evidence to support any of them. Nearly two hundred people were charged with crimes, and some were convicted. Some of those same people were later let out of prison and awarded millions of dollars by the courts when it was determined that the loss of their freedom was just another consequence of the hysteria. With regard to the alleged child victims, it became clear that details were being fed to them through "interrogations" by their families and by police and prosecutors. Rumor and gossip spread like wildfire and before long, the stories were repeated by more and more children and the accusations became more and more bizarre. The whole thing was reminiscent of the Salem Witch trials. It was all nothing more than a massive conspiracy theory legitimated by law enforcement and the media.

Demographics and Contemporary Conspiracy Theories

How is it that all of these seemingly normal people came to believe such an obviously unhinged account of what was (or, in this case, wasn't) happening to their children? This is just one case among many of the phenomenon, perhaps a species of shared psychotic disorder as a result of which delusions spread rapidly throughout a community. It will be useful, then, to say something about the characteristics of a conspiracy theorist, including demographic information and a general psychological profile. It may be the case that conspiratorial thinking is more easily spread among people from similar demographics and with similar psychological profiles. What is the psychological profile of a conspiracy theorist? Is belief in conspiracy theories something that we should be able to predict if we know the other behavioral characteristics of a person?

When we conceive of a conspiracy theorist, it's tempting to think of a socially isolated and ostracized person who spends much of their time at their basement computer. The wild-eyed meme featuring the star of Ancient Aliens with a red thread connecting convoluted lines of thought readily comes to mind. That said, when people are polled about their belief in conspiracy theories, over half of the American population report belief in at least one (Oliver and Wood 2014).

Despite that fact, there is a difference between a person who casually believes in a conspiracy or two on the one hand, and a person who has a strong inclination toward belief in conspiracy theories on the other. In this chapter, we will provide a review of some of the polling and psychological literature related to belief in conspiracy theories. Why are people consistently attracted to conspiracy theories to begin with? What keeps people invested in these theories, even when there is significant evidence that the conspiracy is false? Why are people often willing to give up central features of their practical identities (friends, family, careers) in service of conspiracy theories?

There are challenges to gathering data on this kind of belief. Let's say that we're trying to identify whether a person believes at least one conspiracy theory. We might not get an accurate answer, or an accurate aggregate number because, if a person believes the content of the theory, they are not going to characterize it when asked as "belief in a conspiracy theory." Researchers would likely need to be prepared in advance with a particular set of conspiracy theories and then ask specifically about those (this has recently been done in the case of common contemporary conspiracy theories which we'll explore below). Another challenge has to do with the willingness to self-report; a person may be aware that belief in conspiracy theories is attached to social stigma, so they might not be inclined to respond honestly when polled because they're concerned about how they might be viewed by the pollster. A third challenge has to do with the differences in the type of data that is collected by researchers. Studies are conducted with different questions and interests in mind, so it can be hard to connect data across theories. Nevertheless, we'll see if we can draw some broad conclusions while acknowledging potential shortcomings.

There is some recent polling data available on three major conspiracy theories: that the COVID-19 pandemic is a hoax, that the 2020 election was rigged, and, to the degree to which it is measurable, the family tree of beliefs that is understood as the phenomenon "QAnon." Looking at this data will be somewhat instructive, though there are challenges that we'll identify following the more general discussion.

A survey conducted by Pew Research Center in 2021 indicated that 71 percent of people have heard of the conspiracy theory (or set of conspiracy theories) that maintains that the COVID-19 pandemic is a hoax perpetuated by global elites (Sadeghi 2022). 25 percent of the population polled reported believing that they see some truth in it. Of that 25 percent, 5 percent report belief that the theory is definitely true, while the other 20 percent believes that it is probably true. The poll indicates that educational background is a significant predictive factor in whether a person believes the theory—48 percent of people with a high school diploma or less believe the theory, 38 percent of people who have some college but no degree believe it, followed by 24 percent of people with bachelor's degrees and 15 percent of people with postgraduate degrees.

Political affiliation also seems to make a difference, at least with respect to this particular theory. 34 percent of people who lean Republican believe that the theory is true or probably true, while only 18 percent of Democrats believe the same thing.

Two other determining factors are race and age. People of color, specifically Blacks and Hispanics, are more likely to believe the theory. This may have something to do with trust in the health care system in light of past horrendous treatment by the government. Young people are more likely to believe that COVID-19 is a hoax than are older people. This may have something to do with the fact that the consequences of contracting COVID are both more likely and more severe for older people.

Let's now see if there are similar trends when it comes to the other dominant conspiracy theories for which we have data. A study conducted by the Public Religion Research Institute looked at political affiliation, religion, and news consumption when asking questions about belief in QAnon (PRRI 2021). They asked respondents for level of agreement with the statement, "the government, media, and financial worlds in the US are controlled by a group of Satan-worshipping pedophiles who run a global child sex trafficking operation." At the time of the poll, 15 percent of all Americans expressed agreement with the statement. By political affiliation, 23 percent of Republicans expressed agreement compared with 14 percent of Independents and 8 percent of Democrats. More alarming, perhaps, are the results of the action item question related to the QAnon belief. Respondents were asked to provide their level of agreement with the statement, "Because things have gotten so far off track, true American patriots may have to resort to violence in order to save our country." A small, but not insignificant, 15 percent of the American population agreed that violence might be necessary. By political affiliation, 28 percent of Republicans polled agreed, compared with 13 percent of Independents and 7 percent of Democrats.

The study found that respondents who consumed far right-wing media (OAN and NewsMax) were considerably more likely to believe the QAnon conspiracy. 40 percent of people who regularly watch these news networks believe the theory, by comparison, the results broke down with respect to other news networks in the following way.

Around one in five Americans who do not watch television news (21%) and trust Fox News (18%) agree. Around one in ten Americans or less who trust local news (12%), CNN (11%), broadcast networks such as ABC, CBS, and NBC (8%), public television (7%), and MSNBC (5%) believe this core tenet of QAnon.

There were similar stark differences when it came to the question regarding the need for violence. 42 percent of respondents who consumed far-right media reported that "patriots might have to resort to violence." As far as other respondents are concerned, "Less than one in five Americans who do not watch television news (19 percent) or who trust local news (16 percent) agree, and less than one in ten who trust CNN (9 percent), broadcast news (8 percent), public television (7 percent), or MSNBC (7 percent) agree.

Finally, broadly speaking, religious people were considerably more likely to express agreement with the theory than were non-religious people. This may have something to do with viewing the world in a binary battle between good and evil as we will discuss at length shortly. Among religious denominations, those who were most likely to believe the theory were Protestants of various types. Among Hispanic Protestants, agreement was at 26 percent, among white evangelical Protestants, agreement was at 25 percent, and among other Protestants of color agreement was at 24 percent.

Among other religious denominations or nonbelievers, "Less than one in five Mormon (18%), Hispanic Catholic (16%), Black Protestant (15%), other Christian (14%), non-Christian religious (13%), white Catholic (11%), religiously unaffiliated (11%), white mainline Protestant (10%), and Jewish Americans (8%)" believe the theory.

When it came to the religious question, the only denominations that agreed that violence might be necessary at a rate of 20% or higher were white evangelical Protestants (24 percent) and Mormons (24 percent). The breakdown for other religious denominations or lack of religion were, "Protestants (18%), other Protestants of color (17%), Hispanic Catholics (17%), white Catholics (16%), other Christians (15%), Black Protestants (12%), Hispanic Protestants (12%), religiously unaffiliated Americans (12%), members of other non-Christian religions (11%) and Jewish Americans (6%)."

Finally, the study collected other demographic data for a total overall picture. As a general profile, a person was more likely to believe the QAnon Conspiracy if they were: Republican, Conservative, Religious, consumed far-right media, had no college degree, made less than $50,000 a year, were between eighteen and twenty-nine years old, and lived in a rural area.

Finally, let's look at data related to the theory that the 2020 election was rigged—the theory commonly referred to as "The Big Lie" because its sources are the lies promulgated by former President Trump, his administration, and his supporters. According to a Marist Poll conducted in October, 2021, 62 percent of respondents reported a belief that former President Trump continues to lie about the outcome of the election because he does not like the results (Marist Poll 2021). Among Republicans, however, 75 percent reported that "Trump has a legitimate claim that there were 'real cases of fraud that changed the results.'" The poll did not report information regarding other demographic data.

This data provides something of a general profile about believers in the most prevalent conspiracy theories of the day. There may be reasons not to generalize too much from it. Contemporary political events concern issues with which Republicans might be particularly concerned: the election of a Republican president, a disease that arguably cost that president the election, and a theory about a cabal of pedophiles who were allegedly all Democrats whom Trump was destined to stop. That said, well, Democrats and Independents care about disease and election security too. However, if we were comparing demographics among believers in a fake moon landing, multiple shooters of JFK, and Area 51 aliens, we might get different results. Aside from general demographics and identity categories, there are some psychological characteristics that studies demonstrate tend to be present among people who have a regular disposition to believe conspiracy theories. We'll now turn to a discussion of those characteristics.

Psychological Characteristics and Belief in Conspiracy Theories

Demographics can only tell us so much about a person. From an external perspective, an individual is often defined by their reliable dispositions to behave rather than how they are inclined to self-identify. We learn the character of a person from how they are inclined to think and act. So, regardless of age, race, political affiliation, educational status, socio-economic status and so forth, it will be instructive to look at work that tells us what a person who regularly adopts conspiracy theories into their beliefs is like.

One study suggests that the best predictor of whether a person will engage in conspiratorial thinking is whether they've engaged in conspiratorial thinking before (Swami 2011). Recall the coherentist model that

we discussed in the chapter on epistemology. If conspiracists are working, knowingly or not, under a coherentist model of justification, then, if a paranoid
explanation appealed to them once, it is much more
likely to do so again, especially if the conspiracy is related to a similar topic. If you believe one thing to be
true, and another thing resembles and is consistent
with that first thing, you are likely to be favorably inclined to accept that second thing. For instance, if I believe that raspberries are delicious, and another fruit
resembles raspberries in appearance and smell, I might
be more inclined to believe that fruit is delicious too,
even without tasting it. Most people who believe at
least one conspiracy theory don't adopt a more general
conspiratorial mindset. That said, once a person accepts
one, they are more likely than they were before, and
more likely than other people, to accept more than one.
The poll about QAnon belief discussed above comes to
a similar conclusion: though people in general are not
likely to believe that the QAnon conspiracy theory is
true, those who do believe it are more likely than others
to believe other conspiracy theories as well, including
The Big Lie.

But perhaps this begs the question. It sounds as if
researchers are saying that the people who are most
likely to believe conspiracy theories are people who believe conspiracy theories. Though we think the above
conclusions are more substantive than that tautological
claim, more information about psychological profiles of
such believers would be useful.

Political scientists J. Eric Oliver and Thomas J. Wood
provide one useful framework. They report the results
of the study they conducted between 2006 and 2011.
They identify three common characteristics of conspiratorial thinking. First, believers in conspiracy theories
view the world in such a way that, "they locate the
source of unusual social and political phenomena in un-

seen, intentional, and malevolent forces." Let's look at each of those component parts. First, such believers maintain that the forces involved are "unseen." This contributes to the concerns related to falsifiability that we've discussed in earlier chapters. If the forces that govern the world are, by necessity, unseen, then it's going to be impossible in principle to identify the actors and obtain reliable evidence to expose them and end their machinations. Second, they believe that the actions of such actors are intentional. This self-reporting about beliefs provides further evidence for our earlier claim that conspiracists often avoid attributing the bad things that happen in the world to a chaotic, absurd universe. Instead, world events are orchestrated by people who have the express purpose of seeing to it that things happen in a particular way. Finally, the forces that they believe are in play are malevolent. We're not just dealing with people doing things intentionally, we're dealing with people (or other forces) who are purposefully trying to make the world a worse place and to corrupt the good people in it.

This is related to the second common characteristic of conspiratorial thinking that Oliver and Wood identify. They note that conspiracists "typically interpret political events in terms of a Manichean struggle between good and evil." Manicheanism is a worldview, prevalent across religions and cultures, which maintains that the world is locked in a battle between light and darkness, good and evil. One characteristic that conspiracists have in common, then, is the view that the motivations of the world's actors contribute either to good or to evil and that the evil, demonic actors always strive to "win" the game of history. In light of this finding, the postulation of satanic, ritualistic pedophiles, either in 1980 or 2021 is no surprise. Demonic forces have been appealed to in order to explain world events for centuries. It may also come as no surprise that people who believe conspiracy theories tend to be religious, since religions often take on this binary worldview.

This view that the world is divided into good and evil is combined with both a dislike and a distrust for entities that conspiracists characterize as "elites." Though this term is rarely clearly defined, it tends to include members and leaders of centralized government, certain (but not all) very rich people, academics, and other kinds of researchers. These people are characterized as the evil agents who are dictating the course of world events. Members of these groups don't tend to be portrayed as human beings with complicated psychologies and motivations; instead, they are the dark forces who are bent on destroying the light at all costs. Consider people who are routinely anti-Catholic, anti-Masonic, anti-Communist, anti-Critical Race Theory, anti-Academic, or anti-Antifa. There is always a bogeyman. Sometimes the organization the given theory blames for some particular event exists, but sometimes its very existence is a conspiracy theory.

The third characteristic of this kind of thinking is described as a tendency to believe that "mainstream accounts of political events are a ruse or an attempt to distract the public from a hidden source of power." This is consistent with our general definition of what a conspiracy theory is; in part, it is a claim to have access to information, counter to the prevailing narrative, to which very few or no other people have access.

A 2017 review of research on the psychological characteristics of conspiracists revealed that people with a disposition to routinely believe such theories seek out psychological well-being when forming beliefs (Douglas 2017). In other words, human beings, and conspiracists in particular, form beliefs at least in part on the basis of what makes them feel good. The first sense of psychological well-being is epistemic—it feels bad to be wrong about something. It feels good to be right about something. And, at least for some, it feels *really* good to be the *only one* who is right about something. Regardless of what else is going on in a person's life—whether they are educated, well-liked, well-paid, or well-respected—if they are a member of a small group of people who knows something everyone else is "deceived" about, that feels good, psychologically. It's a source of power and well-being. Conspiracists are prone to engage in confirmation bias—the tendency to only listen to sources that confirm what they were already inclined to believe anyway (to be fair, this is a tendency that most humans have, not just conspiracy theorists, so we should all be on the lookout for similar behaviors in ourselves).

The tendency for these theories to be unfalsifiable helps with this sense of epistemic well-being. If it's impossible, in principle, to prove a theory wrong, then it is also impossible to prove the person who believes it wrong. As a result, they have, from their perspective, a shield against criticism and disagreement and, as a result, protection from psychological discomfort.

The review also focuses on existential sources of psychological wellbeing, in particular the need to feel both safe and in control. In the last chapter, we outlined the ways in which conspiracy theories tend to converge around central human themes. It's no surprise, then, to learn that believers seek out explanations that help them to feel more in control of their circumstances. Consider the coronavirus pandemic. We've been dealing with a virus that has killed millions of people and will no doubt kill many more. It may simply be more psychologically comfortable for some people to believe that the whole thing is a hoax and that there is no need to take any precautions or to follow any safety protocols. The newspapers have been full of accounts provided by medical professionals of interactions with patients on their death bed, finally realizing that they shouldn't have been so brazen in their belief that COVID was a hoax. Alarmingly, others persist in their beliefs about the hoax, refusing to abandon the conspiracy theory right up to the point at which COVID finally kills them. Psychologically, however, deadly disease is frightening and makes people feel out of control. One response some take is simply to deny that it is even happening.

This may be all for nothing. Research suggests that, though a person may seek out a conspiracy theory out of a heightened need for control, they never find the sense of control that they are looking for. Though in some ways tragic, this isn't surprising. If an explanation for an event isn't tethered to reality, that explanation isn't going to provide a person with any causal power in the real world or, if it did, it would be by mere coincidence.

The final source of well-being that the review identifies is social well-being. All humans desire to be a part of groups, even if only small ones. They long for a sense of belonging and to know that they are accepted and appreciated by other members of the group. Social groups are often brought together by sets of shared beliefs and values. People also want to feel that they are in the *right* group—the group that has the largest set of true beliefs and the most ethical or advisable policies. A person gives up more than simply a belief when they abandon a conspiracy theory—they risk giving up their whole social group. In the previous chapter, we discussed concerns about identity and how conspiracy theories tend to cluster around the identities that people have constructed for themselves. Many of these identities have to do with the social relationships that people have taken on. If a person abandons a conspiracy theory, they abandon their support group and a significant portion of who they are, how they conceive of themselves, and their source of self-esteem.

One study, which has been widely reported, suggests that believers in conspiracy theories share characteristics that are common in people with psychopathology (Drinkwater 2012). These characteristics are delusional thinking and an inability or unwillingness to change that thinking when presented with compelling evidence to the contrary. People with psychopathology also exhibit high levels of anxiety and tend to trust their own intuitive judgments over the judgments of others, regardless of the level of expertise of the other in question.

There's much more to investigate when it comes to the psychology of conspiratorial thinking. There will, no doubt, be much more research done in the future, given the impact that such beliefs have had on recent events. It has become clear that a person's beliefs aren't simply an isolated personal matter—they can, and too often do, have devastating real-world consequences. As a result, it may be unethical for us to form beliefs in certain kinds of ways. We turn to this question in the next chapter.

Part III

Conspiracy Theories and Values

12
The Ethics of Belief

The Parkland Shooting

One of America's deadliest school shootings occurred on February 14th 2018. On that day, a shooter walked onto the campus of Marjory Stoneman Douglas High School in Parkland, Florida, carrying an AR-15 semiautomatic assault rifle and a backpack full of magazines. During the shooting spree which lasted about four minutes, the shooter killed fourteen students and three staff members and wounded another seventeen persons. When the shooting stopped, the shooter blended in with the other students and easily fled the scene.

About an hour after the shooting, a suspect, Nikolas Cruz, was taken into police custody. He was eventually charged with seventeen counts of premeditated murder and seventeen counts of attempted murder. Cruz was a former student at Marjory Stoneman Douglas High School and had been expelled for disciplinary reasons. The primary basis for the arrest was a combination of eye witnesses tying Cruz to the crime (including an eye witness who had conversed with Cruz on campus just moments before the shooting began) and surveillance video footage. Magazines found at the crime scene had swastikas and racial slurs carved into them. Cruz had expressed racist, homophobic, antisemitic, and xenophobic views on social media (Wikipedia). He had recently legally purchased an AR-15-style semiautomatic rifle. Cruz was indicted by a grand jury, but at the time of this writing has not been brought to trial (his trial is currently scheduled to begin in February 2022).

On the face of things, this doesn't seem like particularly good fodder for a conspiracy theory. We have a crime, several witnesses, several victims, and a suspect in the crime that is tied to it by past negative events (there appears to be a solid motive in the case). Yet, the Parkland Shooting conspiracy theory has become one of the most active on social media.

The theory began to gain traction as a number of survivors of the shooting started appearing on various news programs and at gun-legislation reform rallies. The survivors saw their celebrity rise in a fairly short time, and there was much speculation that their prowess as spokespersons for sensible gun laws would lead to successful legislation. Cue the conspiracy theorists. Soon all sorts of conspiracy theorists from Alex Jones, and former Republican congressman Jack Kingston, to members of QAnon and other fringe groups declared the event to be fake. These conspiracy theorists denied that the shooting occurred, and claimed that the Parkland survivors were merely crisis actors. (Alex Jones and others propagated similar conspiracy theories about the Sandy Hook Elementary School shootings.) In a similar move, Congresswoman Marjorie Taylor Greene referred to the shooting as a "false flag" planned event, which was not carried out in the way described above, but rather was an event that occurred so that US Speaker of the House of Representatives Nancy Pelosi would have additional support for gun-law reform legislation. The most prominent of the survivors, David Hogg, has been accosted in person by Marjorie Taylor Greene and has been attacked on social media by a number of prominent Republicans and conspiracy theorists, including then President Donald Trump.

Since any good conspiracy theory needs conspirators, it is claimed by some proponents of the Parkland Shooting conspiracy theory that the person who organized the entire event and paid off the crisis actors is George Soros. (How does Soros tie in? One conspiracy theorist suggested that there is no way that a group of kids could pull off the sort of events they were having and publicity they were getting, so it must be Soros. At some point the Soros conspiracy theorists just stopped trying. Sheesh!) Other versions of this conspiracy theory leave out the details, and merely attribute the string-pulling to the "evil global cabal" (either the Illuminati or QAnon's cabal of sex-trafficking Satan-worshipping cannibalistic pedophiles, take your pick). At any rate, none of the versions of this conspiracy theory are particularly well-supported with facts. They amount to nothing more than conjecture by persons who have political motives to discredit the Parkland survivors and those who are expressing sympathy for what they went through.

Believe What You Want to Believe?

Conspiracy theories such as the Parkland Shooting theory raise a further interesting question that we've not yet discussed: is it bad simply to believe conspiracy theories? Suppose that someone believes one of the Parkland Shooting conspiracy theories, but never acts on that belief—they never voice it, nor do they behave differently than they would have had they not believed the theory. Some might argue that their holding the belief was morally neutral. They might claim, as an acquaintance of ours recently did, that "there are no such things as thought crimes." Is belief in conspiracy a "thought crime" or is there really no such thing?

Doxastic Voluntarism

One thing that our question appears to assume is that we have control over our beliefs. Doxastic voluntarism is the view that we have the ability to choose to believe something. Doxastic involuntarism is the rejection of that idea—it's the view that we don't have the ability to choose what we believe. The question of whether we have the ability to choose to believe some particular proposition is one where historically there has not been much consensus, and the issue is far from decided today.

One compelling argument for doxastic involuntarism comes from the eighteenth-century philosopher David Hume. In Section V of his *An Enquiry Concerning Human Understanding,* Hume investigates the difference between entertaining a thought without believing it to be true, and actually believing it to be true. After considering a number of possible candidates, Hume concludes that the difference between believing a thought and just merely entertaining it has to be something more than just adding some additional thought.

We don't simply add the thought "it is true" to some proposition to turn it into a belief. If you are not convinced by this, give the following experiment a try. Take a proposition that you don't believe to be the case, such as that the Parkland shooting didn't actually happen, and add the words "it is true" to that proposition. You now have the proposition "it is true that the Parkland shooting didn't actually happen. Say it out loud. Notice that saying "it is true" first did nothing to make you believe that it is true.

According to Hume, beliefs are feelings. When you have a belief it strikes you in a much stronger way than if you just consider something. So, the thing that's added to mere thoughts to make them beliefs is a certain strong feeling. This sounds right. Just like having a desire, or a fear, or any other mental state that you can really feel, there is more to it than just thinking these things. The difference between just imagining some state of affairs and actually believing that it is the case is analogous to the difference between being hungry and just thinking about being hungry or the difference between being in pain and just thinking about being in pain.

If Hume is correct about this, belief is not a matter of choice. Just like being in pain or being hungry is not a matter of choice, we can't merely choose to believe something. If belief is not a choice, then one might be tempted to conclude that the question of whether it is morally objectionable to accept conspiracy theories is moot. Here one might employ another of Hume's principles: ought implies can. The idea being that if one is going to be deemed morally praiseworthy or blameworthy for holding a particular belief, such as the belief that one of the Parkland Shooting conspiracy theories is true, then it must be the case that one plays some role in accepting that belief.

We accept Hume's principle, but want to resist the temptation to apply it to belief in conspiracy theories. While we are inclined to accept Hume's argument that doxastic involuntarism is true, we reject the idea that doxastic involuntarism gets one off the hook, so to speak, morally for their beliefs. This is because one still wields quite a bit of influence over their beliefs, even if final acceptance of beliefs is a process over which one has no control.

To see this, consider a simple example. Suppose that Lori has virtually no beliefs about the Peloponnesian War (save for a handful of really general beliefs such as that it took place). If Lori were to read an authoritative text on the subject, certainly she would come away with all sorts of beliefs about it. For example, she would likely come to believe things such as, that it was a war between Athens and Sparta, that Sparta won, and so forth. So even though Lori didn't actually choose to believe any of those things, the fact that she chose to read the authoritative text played a causal role in her forming the beliefs that she formed. This is where the "ought implies can" principle comes into play—it is only reasonable to say that a person ought to do something when they realistically can do that thing. Whether to investigate the matter was up to her, it was her choice, and as a result she is morally responsible for doing so, even if, in this case, her actions turn out to be pretty neutral, morally speaking.

Let's apply this to belief in conspiracy theories. If doxastic involuntarism is true, we don't have control over our beliefs, but we do have control over the evidence we access. Finding weak evidence should motivate us to find more evidence, or to investigate further, and if those things are not possible, we should reflect on the lack of evidence we have. Each of these things greatly decreases the likelihood that we will end up believing conspiracy theories that fail to be supported by compelling evidence.

The payoff is that it is an open question whether doxastic voluntarism or doxastic involuntarism is the case, but either way, we may be at least partially morally responsible for those things that we believe.

W.K. Clifford on Belief

We can now return to our question about whether mere belief in conspiracy theories (in cases where we don't act upon those beliefs) is morally objectionable. One response comes from the nineteenth-century philosopher W.K. Clifford.

In his philosophical artile, "The Ethics of Belief," Clifford argues that we should never believe anything without being in possession of a sufficient amount of evidence. It is clear that Clifford is embracing doxastic voluntarism, but we can take his admonition to include seeking more evidence in the event that doxastic involuntarism is correct. Let's briefly canvass his arguments.

Clifford begins by telling a story about a man who owned a ship that was not all that well constructed to begin with and had become pretty rundown over the years. The ship was about to set sail across the ocean carrying a group of travelers who were emigrating from their homeland. The owner of the ship had a number of doubts about whether it was seaworthy. The thought that it might not be made him fairly despondent; he considered spending a lot of money to have it overhauled and refitted. However, before the ship went to sea, he managed to put his doubts to rest by focusing on the fact that the ship had always made it safely back from previous voyages and "reasoned" that it would likely do so this time, as well. Eventually he came to believe that it was thoroughly safe and seaworthy. The ship sank. Everyone aboard died.

This raises the question: was the ship's owner morally responsible for the deaths of the emigrants? Do we want to let him off the hook because he sincerely believed that the ship was seaworthy? Presumably we don't, because he wasn't justified in his belief. Clifford's position is that the ship owner didn't have the right to believe that the ship was seaworthy on the basis of the evidence that was available to him.

Importantly, the ship owner believed that the ship was seaworthy, but he didn't come to his belief by considering all the relevant details (such as the physical state of the ship); rather he came to hold this belief by repressing his doubts and applying a weak inductive argument (Since the ship made it in the past, it will make it this time).

By contrast, if an expert had told the ship-owner that the ship was seaworthy, he would not have been morally responsible; he would have done his moral duty. So, the important thing, at least morally speaking, is to be justified (to have good evidence) for our beliefs. Even if the ship hadn't sunk, Clifford maintains that the shipowner would have been morally culpable for his belief. Just because the owner turned out to be right doesn't mean that his actions (or inaction in this case) were permissible. The ship-owner just got lucky, but he still failed to act in a morally responsible way. Clifford points out that once an action is done it is right or wrong forever regardless of how that action turns out. (Consequentialists, of course, deny this claim).

So, according to Clifford, the ship-owner was morally culpable the moment the passengers left the dock (whether the ship turned out to be seaworthy or not). Again, this is because he did not have a good reason for his belief that the ship was seaworthy.

Clifford considers a second case, one that happens to share certain salient features with the Parkland Shooting conspiracy theory. Suppose that there is an island where some of the inhabitants have religious views that are vastly different from most other persons on the island. They are members of some kind of religious cult. Also suppose that they are thought by the others on the island to use unfair means to teach their religious principles to children: they sometimes kidnap children from their parents, they don't let children interact with their friends and families, and so on. Now there's no hard evidence that these persons actually engage in these practices; it is just a rumor that is making its way around. Upon hearing this rumor, a group of men formed a society whose purpose was to arouse public anger towards the cult. The society published accusations against high-ranking individuals in the cult and did whatever they could to disrupt their work and personal lives. A commission was formed to look into the matter. The commission drew two conclusions: that the cult members were innocent of the accusations, and that the society could have easily determined that the cult members were innocent with just a minimum of investigation.

Again, the fact that they sincerely believed that the accusations they were making were true doesn't appear to be morally relevant. Clifford points out that even though they sincerely believed in the charges that they were levying, they didn't have the right to believe them.

This case is different from the ship-owner case in that these people aren't repressing beliefs or ignoring evidence, but morally speaking they are still guilty of the same crime: they are accepting beliefs on the basis of insufficient evidence.

Again, had their accusations turned out to be true, the society members would not have been vindicated. The moral issue in this case doesn't have anything to do with whether their accusations are true. Rather, it has to do with whether the accusations are fair or just. The only thing that could make them fair or just is strong supporting evidence. So, either way, since the society was acting on rumors (as opposed to facts supported by evidence), they committed an injustice against the cult members. Clifford famously concludes, "It is wrong, always, everywhere, and for anyone, to believe anything upon insufficient evidence."

Belief versus Action

One natural objection to raise to Clifford's conclusion is that in both the ship owner case and in the cult rumor case, the thing that was morally objectionable was not *holding* the belief, but rather, it was *acting* on the belief. Clifford agrees that even if my belief is fixed, I can control my action, and I have duties to *act* in certain ways (for example to have my ship checked before sending it on a long voyage) even if I don't believe there is anything wrong. But he thinks the original judgment still stands: if the belief was gotten illegitimately—if it came about without relying on good evidence—then the person who holds the belief is open to moral criticism, and has failed in his or her duty. This is because belief is not simply disconnected from action. To hold a belief involves having some tendency to act in certain ways. And if I hold a strong belief without evidence, it will cloud my judgment when I try to carry out the duty of investigating facts carefully. We see ample evidence in the case of conspiracy theories. Initial belief in the original version of the QAnon conspiracy theory has made belief in the additions to the theory easier for many QAnon members. Recall that the Big Lie was presented with virtually no evidence, but those disposed to accept other conspiracy theories were among the first to accept the Big Lie.

Clifford also points out that belief isn't just a private matter. On his view, *any* case of believing for faulty reasons has the potential to infect and corrupt the system of belief upon which we all depend. And any such act weakens our self-control and our critical faculties. And carelessness about the evidence leads eventually to carelessness about the truth itself. We'll discuss this at greater length in the next chapter.

The consequences for belief in conspiracy theories should be clear: if Clifford is right, then believing them without sufficient evidence is wrong. Even the most innocuous cases of such beliefs unsupported by sufficient evidence have a morally objectionable component.

13
Conspiracy Theories and Human Virtues

The (Failed) Resurrection of JFK Jr.

On a rainy late fall day in 2021, hundreds of people congregated at the site at which President John Fitzgerald Kennedy was assassinated. The purpose of the gathering was not something that people could have easily anticipated. The attendees, most of them QAnon adherents, thought that this was the big moment they had been waiting for—the moment Trump would be put back in control. They believed that the signs indicated that on this day, November 2nd 2021, JFK Jr., the son of the murdered president, would reveal that he did not really die in an airplane accident in 1999. He would emerge from hiding and become Trump's new vice president (Pence having been disgraced for refusing to overturn the results of an election).

Never mind that Trump and Kennedy differed significantly in their politics, some members of the congregation believed that the event would precipitate the reinstatement of Trump, others believed that JFK Jr. would announce himself to be Trump's running mate in the 2024 election, and still others claimed that JFK Jr. was coming out of hiding to, by some authority, declare Trump to be king of the country. This view is perceived as fringe, even among QAnon adherents, but there were enough people gathered to generate significant concern about the extent to which people have lost their grip on reality.

Conspiracies and Character Development

What we believe shapes who we are. The social circles of which we are a part, the media we consume, the ideas we entertain, and those we rule out as unimportant or unworthy of our time serve to carve out our personalities. Most of us probably desire to become the best version of ourselves. Doing so requires engaging in the hard work it takes to develop certain kinds of virtues and to avoid vice. When evaluating what our orientation to conspiracy theories should be, there are several approaches that we can take, and we'll employ more than one of them in this book. One approach to ethics is to focus on character—what actions do we need to take to excel in the development of good character? What actions do we need to avoid? Should we spend our time pursuing conspiracy theories? Does doing so harm our development?

In conversations with students, at least prior to the pandemic and the events surrounding the 2020 election, they would report to us that belief in conspiracy theories struck them as harmless and perhaps even, in certain cases, fun. This isn't that surprising. After all, many of us grew up with movies and television that lauded the virtues of, shall we say, "imaginative" belief formation processes. Those who tuned in religiously to watch *The X-Files* in the 1990s would be lying if they didn't admit that there was something admirable about the character of Fox Mulder. He was, after all, the show's main protagonist. Many even had a copy of his iconic "I Want to Believe" poster hanging in their bedrooms. What, if anything, is heroic about such a mantra? After all, we should believe things, not because we "want" to believe them, but because we have good evidence for thinking that they are true.

What made us think of Fox Mulder as a good guy? Why was he the hero for blatantly violating Ockham's Razor every time he had the opportunity? We'll put aside the fact that, by stipulation, Mulder turned out to be right that aliens, vampires, psychics, and other mythical creatures actually existed in the universe of the show. Why did we cheer for Mulder, knowing full well that he embraced the explanations that were least likely to be the right ones (at least in our world)? There are many Mulder types walking the American streets, wearing their belief in conspiracy theories like a badge of honor.

Even Mulder wasn't conspiratorial enough for his friends who ran the conspiracy theory rag, *The Lone Gunman* (who eventually got their own, ultimately unsuccessful spinoff). One noteworthy difference between the events that took place on *The X-Files* or *The Lone Gunman* on the one hand, and conspiracy theories in real life on the other, is that the conspiratorial plotlines on the television programs were always ultimately either confirmed or disconfirmed by relevant evidence. That is, unlike so many of the conspiracy theories entertained in real life today, they were, in principle, falsifiable.

Fewer people are watching *The X-Files* these days, but plenty of people watch YouTube videos where conspiracy theories spread like a virus. This phenomenon is exacerbated by the fact that people who view them then share them on social media. As the Facebook whistleblower revealed—social media has a problem with promulgating misinformation and they know it.

We like to suspend our disbelief. Consider how you might feel going to a scary movie or a haunted house at Halloween. You may not believe in ghosts, vampires, or Freddy Krueger, but you allow yourself to put aside your ordinary beliefs for a moment to participate in the fun. You get all of the endorphins associated with feeling fear, but (if everything is done responsibly) no one gets hurt.

One significant difference between the situation we've just described and the ways in which conspiracy theories play out in the real world is that sometimes, as a result of the proliferation of inaccurate information, people *do* get hurt. In other words, when we believe conspiracy theories in the real world, we aren't *simply* suspending our disbelief. People share beliefs; a religious group may share in common the same set of beliefs about God, morality, and the afterlife. A political party may all have the same beliefs about social policy. On *some* level, however, beliefs are private things. Beliefs are contained in individual minds, and individuals knowingly or unknowingly engage in behaviors that produce and sustain them. They often think they should be applauded for doing so.

Conspiracist—Loud and Proud

We have an acquaintance who's a very proud conspiracy theorist and expresses that pride regularly on social media. It's difficult to think of a conspiracy theory that she does not endorse; she has certainly expressed confidence in each of the most popular ones over the years. This acquaintance reports that she frequently gets private messages thanking her for her bravery— she is saying what they wish they could say but are too scared to express for fear of the potential repercussions. Paying attention to this person's Facebook page is like rubbernecking a car accident—it's ugly, but it's such a mess that it is hard to look away. One thing's clear— this person thinks she is a major community hero for holding the conspiratorial line. Her belief in these theories—her confidence that she is one of the select few people who knows things about the universe that no one else knows (even though she tells everyone about them compulsively and ceaselessly on social media)— is clearly a significant source of her self-esteem. Many of her former friends have found her beliefs, increasingly racist and antisemitic, too much to handle, even casually. She frequently comments on the friends she has lost, but, consistent with a Manichean way of looking at the world, she insists that she has simply shed friends who turned out to be part of the darkness and found others on the side of the light. Even after shedding friends who struggled with her world view, she has more friends than ever before because her set of conspiracy theorists have become a tribe.

In light of all of the self-reports, it would be silly not to take such a person at their word when they claim that they feel superior to others because of their commitment to belief in conspiracy theories. What might they mean by that? What might such people view the character virtues connected to this kind of world-view to be?

Before we begin, we want to be clear that we are not suggesting that conspiracy theorists are devoid of all virtue. We have friends and family members who are conspiracy theorists, but are otherwise lovely people. In the following sections, when we discuss virtue and vice, we are discussing traits only as they apply to belief in conspiracy theories and not as they might be applied to other features of a person's life, traits, and behaviors.

Courage

One of the virtues that Aristotle, the world's most famous virtue theorist, discusses at the greatest length is courage. When we identify courage correctly, this does, indeed, seem to be something that most people value. Virtue theory often involves pointing to moral exemplars who excel at particular virtues. Consider the advice one often hears to reflect upon the question "What would Jesus do?" The idea here is to look at someone like Jesus—a figure who many consider to be a perfect moral agent—and try to exhibit similar behavior and virtuous traits in one's own life.

We tend to admire the courage of soldiers who left their comfortable lives at home to travel to Europe to fight the Nazis. We admire the courage of a figure like Martin Luther King Jr., who was willing to go to jail multiple times and to face death threats, even up to a successful attempt, in order to create a more just world. We admire the courage of a close friend diagnosed with a terminal illness who, knowing that he only had months left on Earth, lived his live with love and good humor until the end. Yes, it does seem that courage is something to be admired. It's a trait that each of us should try to take on, if possible.

It's also true that sometimes courage does have to do with maintaining a belief or position, even when doing so may be harmful in certain ways. History casts Thomas More and Catherine of Aragon as heroes for refusing to denounce their Catholic beliefs even under extreme pressure (and, in More's case, execution) by King Henry VIII. We tend to think of Martin Luther as courageous for writing his *95 Theses* expressing disagreement with the Catholic Church and maintaining his commitment to them, even under extreme pressure. In modern culture, many value the courage of Greta Thunberg for passionately advocating climate change action even in the face of threats and Malala Yousafzai for her commitment to action in support of education for women and girls in Pakistan, even after she was shot in the head for doing so. So, yes, it can also be virtuous to believe things and to stand up for those beliefs.

The question is: is courage being exhibited when a person believes a conspiracy theory or a collection of conspiracy theories? There are a few different approaches we can take. First, we typically don't include just any behavior that flies in the face of risk in the category of what we are inclined to call courageous. One reason may be that it is not courageous to believe in something that is just clearly, demonstratively wrong. Imagine that a child goes to school and takes a test. The child's parents have tried to pass along to that child the value of becoming well educated and of respecting the normative connection between evidence and belief. One day, the child comes home and shows their parents the results of an essay test they needed to take. The child failed the test because they answered every question with what they were inclined to believe before taking the class and learning the material. Should the parents praise the child's bravery, or should they insist that the child tries harder next time to appreciate the evidence presented in class?

Similarly, suppose a person performs a bad inference. Imagine that they commit what philosophers call a "formal fallacy," reasoning in the following way:

Premise One: If P then Q.

Premise Two: Not P.

Conclusion: Therefore, not Q.

A colleague of ours refers to this inferential move as "modus bogus" (a pun on similarly looking valid argument forms called "modus ponens" and "modus tollens"). Modus bogus is always invalid. If a person reasons in this way, the appropriate response is not to praise their "courage" for making the inference that no one else dared to make—it is to point out to them that their inference is invalid.

The same is true when a person commits an informal fallacy. When someone says "Don't accept the conclusion of Sam's argument, he once wet the bed in the third grade!" we don't (or, in any case, shouldn't) praise that person's "courage" for bringing up that memory about Sam. It's an attack on the person rather than on the merits of the argument, and as a result it is irrelevant to the question of whether we ought to accept the conclusion of the argument based on the premises provided.

Second, we don't typically deem behavior to be brave if it involves unnecessary self-harm. We don't praise someone's virtue when they needlessly stick their hand in a pot of boiling water. Sure, doing so involves risk, but it is a type of risk the person shouldn't have attempted to take on. A person is a fool for having done so, not someone we should emulate. Nothing was gained and there were some pretty significant losses in the form of serious burns.

Third, and perhaps more importantly, we don't typically deem behavior to be courageous if it involves unnecessary harms to others. We don't view a person as praiseworthy for dumping a pot of boiling water over their child. Doing so involved risk, but it was *certainly* not a risk to which others should have been exposed. There are truly innocent victims in all of this. We'll talk more about other types of harms in the next chapter, so here we'll focus our attention primarily on the way this kind of behavior harms character development. In the previous chapter, we discussed the view called "doxastic voluntarism" and considered Hume's arguments that the position is false. You may have yet to make up your mind on this debate. If we have any control over our beliefs, however, we likely have much less control when we are young and influenced heavily by the people around us whom we have little choice but to trust. Insofar as people infect (figuratively, but also sometimes literally) impressionable people with dangerous misinformation when they spread conspiracy theories, the harm to character doesn't only happen to the original believer.

One of Aristotle's important insights is that virtuous and vicious characters don't just spring into existence—they are habituated. We become who we become through watching and copying others. The development of virtue is a social enterprise. When a person harms their own character by employing unreliable methods of belief production, they also harm the characters of the impressionable people around them.

There are two types of responses we'd offer to the conspiracy theorist who claims that they are exhibiting the virtue of courage by believing what they believe. There are two ways we might understand the value of the virtue of courage. First, we might think that courage is intrinsically valuable—that it always has value whenever it occurs because courage is simply a valuable thing. If this is the case, then the appropriate diagnosis of what is going on in the case of conspiracy theorists is that they have simply misunderstood what courage is—the behavior that they are engaging in is not courage at all, and there is nothing valuable about it. The alternative is that courage is instrumentally valuable rather than intrinsically valuable, that is, it is valuable only insofar as it brings about other states of affairs of value. If this is the case, belief in conspiracy theories may turn out to be courageous, but not in a way that reliably brings about any positive consequences. The behavior of the conspiracy theorist fares poorly on this account as well.

It's also worth noting that, if we take the psychological literature on conspiracy theories to be instructive, a person is likely not exhibiting courage when they believe theories counter to the best evidence. If courage is a sort of strength of mind, resilient in the face of reasonable risk, this is precisely *not* how the conspiracist behaves. Instead, they take on the belief (knowingly or unknowingly) that is most conducive to their own psychological well-being. As we have seen, if a situation has the potential to frighten them, a common strategy is to commit the informal fallacy of *denial*—to behave as if wanting some fact or state of affairs to be false is sufficient for making it false. On the flip side, the person might be committing the *wishful thinking* fallacy: making the misguided move from wanting something to be true to concluding that it must be true. True courage with regard to belief would be following the best available evidence and then believing the conclusion that the best evidence supports, at least until better evidence comes along.

Self-Reliance

Another virtue that conspiracy theorists frequently report having is something like self-reliance—they don't rely on the research and reasoning of other people, they do their own research and draw their own conclusions! Again, this might be a virtuous trait when all of the concepts involved are being defined correctly. We generally try to cultivate self-reliance in our children. For example, we tell them that they will need to take out the garbage, do dishes, and cook meals for themselves when they are older, so they need to learn such skills when they are young so that they'll be effective at them later. Perhaps even more importantly, we try to guide them toward general strong problem-solving skills so that they'll be in a position to solve a wide-range of problems as adults.

That said, most enterprises that we set out to accomplish in life are things we don't do alone and are things that are probably best not done alone. For example, when I cook a meal, I do not rely exclusively on my own knowledge. I depend, critically, on the knowledge people employed to create stoves, pots and pans, knives, utensils, and so forth. I rely on supermarkets to sell fruits and vegetables that are not poisonous. I often rely on recipes which help me to discern delicious flavor combinations from those which may taste disgusting. If we tried to do things truly, entirely on our own, we might make impressive gains by primitive standards, but we wouldn't do that well by contemporary standards. Every day, we take for granted that virtually everything we are able to accomplish depends in some way on the labor of others.

Imagine Ug is a member of a primitive society of human beings. In the environment in which Ug lives, clean drinking water is hard to come by. The community works together, using a method of trial and error, and develops a way to reliably clean the drinking water at a particular local water source. One day, Ug wonders off into the woods and drinks water from a previously unknown source. He gets very sick. If Ug claims to the members of his social group that he was simply being "self-reliant" the appropriate response would be "No, you were being foolish. Knowledge derived from careful attention and experience was available, and you willfully ignored it."

Like virtue, knowledge is often a social activity. The wide range of things that we know, we know because we've spent a significant chunk of human history developing reliable ways to arrive at beliefs. That said, there are models in the history of philosophy who may serve to empower this individualistic way of thinking about knowledge and self-reliance. For example, in Descartes's *Meditations on First Philosophy* he sets off to construct an epistemology as a solo mission. He sits by his fire and thinks about his own beliefs. He realizes that some of them have been false. He decides to construct skeptical hypotheses to cast as many of them into doubt as possible and see what remains. At the end of the day, he is left with two beliefs that he takes to be immune from doubt "I think" and "I exist." From these two basic, foundational beliefs, he thinks he can derive all the rest of his knowledge. It's difficult to imagine an epistemology that is more self-reliant than that.

This isn't the right place to explore whether Descartes's project is successful (we don't think it is). It's clear that what we know does not, largely, come from introspection. Much of what we find upon introspection is confabulated rather than found (for example our "reasons for action" are often rationalizations after the fact). Instead, most of what we know, we know from some form of testimony or reliance on the hard work of others. Still, the conspiracy theorists might point out that many people accept, without hesitation or critical evaluation, the explanations and justifications offered by authority figures. Conspiracy theorists insist that they are more on guard against fallacious appeals to authority than others. After all, explanations do not become accurate simply because the president, a member of Congress, or some committee chair claims that the explanation is accurate. Many conspiracy theorists, then, think that they are employing more strenuous critical thinking methods than the average citizen. In fact, recent studies suggest that conspiracy theorists tend to be poor critical thinkers (Elder 2021).

Conspiracy theorists may also claim that they attend to evidence in greater detail than others who do not consider such theories. They may claim that they pay more attention to counter evidence than the casual observer and that, as such, they are in a better position to know what really happened. They may argue further that conspiracies *do happen* sometimes, and if they do happen on occasion, it is an open possibility that they are occurring in any particular case. (It is worth noting that Descartes considered an argument very similar to

this while considering skeptical arguments, but ultimately rejected it because he recognized that the conclusion simply did not follow.)

The conspiracist has introduced skeptical counter-
possibilities to make us question whether what we
think is the best evidence really *is* the best evidence.
Though they may have provided some reason for doubt,
some reason for doubt shouldn't shake us from our commitment to a best evidence standard. The fact that we
might have *some* reason for doubt certainly doesn't
mean that we should revert to a wacky, racist, or antisemitic explanation instead! We should just be very cir-
cumspect in our views about evidence and the kinds of
things that might demonstrate that our beliefs are
false. The best evidence is frequently found via the reliable methods we've worked out together. Self-reliance,
when it is virtuous, doesn't demand that we disconnect
from one another completely when forming our beliefs.

Consider two people, Tom and Jane. Tom believes it
will rain tomorrow because Tom heard it from his local
meteorologist. Jane also believes that it will rain tomorrow, but she believes it because her psychic mentioned
it in their weekly session. Tom's belief was formed using
a more reliable method—meteorology, which may be far
from perfect, but is far more truth-conducive than the
speculations of a "psychic."

Imagine that Sarah rides her bike to work. Jane tells
her that she had better take the bus because it's going
to snow. Sarah, not knowing that Jane gets much of her
information from her psychic, makes the reasonable as-
sumption that her friend was not just pulling beliefs
from the clear blue sky, but instead has some compelling evidence for them. If it doesn't snow and Sarah
learns that Jane told her it would as result of the speculations of a psychic, she would be rightly annoyed with
Jane for causing her to miss out on her biking exercise
that day. Jane may think that she was being "self-

reliant" in choosing sources that may be inconsistent with the sources that have a demonstrated reliability, but she's really just being foolish, and she is harming people by spreading misinformation.

Again, self-reliance can be a good thing when understood in certain ways. However, charging out on one's own and avoiding all of the careful labor done by others to keep human beings safe and to help them arrive at true belief is not praiseworthy. That's not to say that non-standard sources should never be given a voice. They just shouldn't be given a voice at the exclusion of all others. The reliability of the advice of the psychic should be measured against the reliability of the advice of the meteorologist.

Loyalty

Conspiracy theorists do not all appeal to loyalty as a virtue that motivates or characterizes their beliefs, but some do, especially when it comes to the set of conspiracy theories that have become prominent in recent years. Loyalty to Trump causes people to believe, unwaveringly: that Trump actually won the election and that Biden's apparent "victory" was really the result of fraud, that COVID-19 is either a hoax or not as bad as it appeared, that the insurrection at the Capitol that took place on January 6th 2021 was really perpetrated by "Antifa," and so on. This kind of commitment come what may to the narrative, talking points, or propaganda of a leader is not uncommon, people reacted or continue to react to Putin, Hitler, Stalin, Caesar, and other noteworthy (or notorious) leaders in the same way. People no doubt respond this way to Biden, Clinton, Obama, and others as well. Again, people want to believe they are on the right side and they want to act in ways that members of their own in-group would find acceptable, those considerations might govern their beliefs more than they realize. They may categorize this tendency to follow their leader or their group as "loyalty."

Is loyalty a virtue? We can think of cases in which it seems desirable. Acting in ways that consistently demonstrate respect toward friends and family is a good thing. Keeping promises you have made to everyone, but particularly those that matter to you, is important. That said, neither of the cases we've described are cases of unqualified goods. We owe all human beings respect insofar as they have inherent worth and dignity. The question of whether we should treat someone with respect in other ways such as following their example and believing what they believe will depend on the nature of their example and what it is that they believe. If we're loyal no matter the circumstances, we run the risk of becoming complicit in really terrible beliefs and behaviors. We can continue to love friends and family even if they engage in bad actions or entertain harmful or poorly formed beliefs. Love and loyalty are not the same thing.

A person might be a conspiracy theorist out of a sense of loyalty, not to a person, but to a country or another institution such as a religion or a political party. The conspiracies that they believe might then be propaganda that the leaders want them to believe. This was certainly true with the propaganda machine headed by Joseph Goebbels in Nazi Germany—citizens were encouraged to believe whatever it was in the Nazi Party's interest for them to believe. There was no lie too great. Even all of these years after the fact, there are people across the world who deny the existence of the Holocaust. Here in the United States, it's not uncommon to hear people deny the severity of treatment of Native Americans since colonization or the severity and continued injustice caused by the many years that the United States allowed for slavery. A person might feel inclined to ignore or deny the existence of all these things out of a sense of loyalty to their country.

If loyalty is a virtue, then we simply don't see it in these examples. We think it might be more reasonable to say that loyalty isn't really a virtue at all if what it entails is remaining true and committed to something no matter what, regardless of whether such commitment is warranted.

Faith

Some people claim to believe conspiracy theories on the basis of religious faith. This is a strange phenomenon, but one that is well documented and is consistent with the demographic and psychological information that we discussed in Chapter 11. Many people believe that certain circumstances call for a suspension of the rules of logic or critical thinking—we take some things on faith alone. This is the position of some religious believers when it comes to issues like survival after death, belief in an afterlife, cosmic justice, and resurrection.

Søren Kierkegaard believed that true commitment to the religious life sometimes requires belief in the absurd, where the absurd is that which is contrary to reason. Sometimes religious demands call on us to abandon reason. He uses the example of Abraham and Isaac from the Bible; Abraham must suspend his reason that tells him that a just God would never require him to kill his own son and must carry through with God's request anyway—he must engage in what Kierkegaard calls a "Teleological Suspension of the Ethical," abandoning what reason might dictate in favor of what the religious life calls for. If you can't do this, you aren't truly living the religious life.

It's beyond the scope of this book (and the scope of our desires) to comment on faith as a virtue when it comes to religious matters. That is a conversation for a different context. What we will say is that we do not think that it is reasonable or religiously faithful to connect Trump, QAnon, or the 2020 election, to belief in God. A person is no more Godly or pious if they believe in conspiracy theories than if they don't. Indeed, it would really be strange if the God of the universe strongly desired the creatures made in his image to believe theories, say, about Satanic, pedophilic, cannibals that are completely unhinged from any available evidence.

However, some conspiracists are appealing to a different kind of faith—not a faith in God, but a faith in their own gut. They harbor a strong belief that the best course of action is just to believe whatever feels to them like it must be true. We maintain that this is not a virtue at all, it is an unreliable method of belief formation. It is consistent with the findings included in Chapter 11 which suggest that conspiracists have characteristics in common with people with psychopathology—in this case the belief that their gut instincts are more trustworthy or reliable than another person's, or, more importantly a well-informed group's, hard-won evidence.

Evidence that should count as disconfirming when it comes to the reliability of this "gut instinct" method of forming beliefs is the fact that different people's guts tell them different things. These things are frequently incompatible. If one person's "gut" tells them that Trump really won the election and another person's "gut" tells them that he didn't, which one is right? Which one should any given person believe? Some people think the answer is, just, obviously "my own." This response, much like Descartes's, makes the self the grounding point for all knowledge in the universe—the hubris is impressive. It does so without any justification, there is no reason to believe that forming beliefs in this way will be any more reliable than any other method of belief formation—in fact, there's good reason to think it will be less reliable, since methods we've developed together have more of an established track record. For this reason, we don't think faith of this type is a virtue at all.

Character, Truthfulness, and Gullibility

In the last chapter, we discussed W.K. Clifford's argument for the claim that one ought never to believe anything on the basis of insufficient evidence. We provided some of the arguments for how doing so harms society as part of that discussion. Clifford also has something to say about character development. He says,

> If I steal money from any person, there may be no harm done from the mere transfer of possession; he may not feel the loss, or it may prevent him from using the money badly. But I cannot help doing this great wrong towards Man, that I make myself dishonest. What hurts society is not that it should lose its property, but that it should become a den of thieves, for then it must cease to be society. This is why we ought not to do evil, that good may come; for at any rate this great evil has come, that we have done evil and are made wicked thereby. In like manner, if I let myself believe anything on insufficient evidence, there may be no great harm done by the mere belief; it may be true after all, or I may never have occasion to exhibit it in outward acts. But I cannot help doing this great wrong towards Man, that I make myself credulous. The danger to society is not merely that it should believe wrong things, though that is great enough; but that it should become credulous, and lose the habit of testing things and inquiring into them; for then it must sink back into savagery.

When we believe things because our gut tells us to, or because they feel good to us, or because our in-group believes them, we harm both our moral and our epistemic characters. We become the kinds of creatures that don't value evidence. We find ourselves, by our own activity, unmoored from reality—we enter a "post truth" society. When we habituate ourselves to believe in this way, we lose our reliability, we become more gullible and less discerning. In the next chapter, we'll discuss the consequences of that at greater length.

14
What's at Stake When We Believe Conspiracy Theories?

Fall Guys and Evidence Planting

In 2015, Netflix released the first season of the docu-series *Making a Murderer*. The series follows the story of convicted murderer Steven Avery. Avery's case is noteworthy because, in 1985, he was wrongfully convicted for the rape and attempted murder of Penny Beernsten. The Innocence Project used DNA technology that did not exist at the time at which Avery was convicted to prove that he was innocent and that a different man had committed the crime. Avery was released in 2003 and subsequently filed a $36 million lawsuit for unlawful conviction against Manitowoc County, among others.

In 2005, photographer Teresa Holbach went missing. Her most recent scheduled appointment was to photograph a van at Avery's home for his family business, Avery's Auto Salvage. Charred fragments of Holbach's bones were later found in a fire pit on Steven Avery's property. Avery and his young cousin, Brendan Dassey were convicted of Holbach's murder.

As a docu-series, *Making a Murderer* was widely successful. Many viewers were left with the impression that the evidence against Avery was planted and that the police misconduct was motivated, at least in part, by a desire to affect the outcome of the impending lawsuit. As a result, there are countless citizens in the United States who believe the conspiracy theory that Avery was convicted by a police department that simply wanted him out of the way and fabricated any evidence that they needed to in order to make it appear that he was guilty.

You might think that the way that the details were presented in the first season of *Making a Murderer* was morally questionable. The producers of the series left many details out, including the fact that a sample of Steven Avery's touch DNA was found on the hood latch of Teresa Holbach's car. Many viewers of the Netflix series became very personally invested in the case, going so far as to write letters to Avery, Dassey, and the law enforcement officials they hold responsible for what they view as the wrongful conviction of two innocent men.

This is just one among many, many criminal cases in which a defendant has claimed conspiracy on the part of the police force. The most famous case of this type was the trial and acquittal of O.J. Simpson. Simpson was accused of stabbing and killing his ex-wife Nicole Brown and her acquaintance Ron Goldman. Police found Simpson's DNA at the crime scene, mixed with blood samples from both Brown and Goldman. Samples from all three individuals were found in Simpson's Bronco, on a bloody glove found behind Simpson's home, and Simpson's and Brown's DNA were found on a pair of socks in Simpson's bedroom. There was also significant hair and fiber evidence connecting Simpson to the scene.

Simpson's lawyers claimed that his arrest and prosecution was the result of a massive racist conspiracy within the police force. In the immediate aftermath of the L.A. riots in response to beating of Rodney King, the public was already disinclined to believe the police. At the end of the trial, one of Simpson's defense attorneys famously advised the jury, in reference to the bloody glove found at the crime scene "If it doesn't fit, you must acquit." Acquittal is exactly what took place.

In the late 1970s and early 1980s, the dead bodies of twenty-eight African American boys and young men were discovered in various locations across Atlanta. Most of the victims had been strangled to death. In May of 1981, Atlanta police were conducting surveillance of the Chattahoochee River, a location where the bodies of victims had been dumped in the past. When police officers heard a splash coming from under a bridge, they apprehended a man whose car was stopped on that bridge, twenty-three-year-old African American man Wayne Williams. The body of twenty-eight-year-old Nathaniel Cater later washed up in that approximate location. Many of the bodies were united by sets of common, extremely unusual hairs and fibers which were ultimately linked to Williams's home and car.

Despite the evidence pointing in the direction of Williams, many people in the community, particularly African American members of the community, believed that Williams was not really the child murderer. He was only ever convicted for the murder of Cater (who was not a child). The disappearance of the boys and young men was a scandal in Atlanta—many members of the Black community believed that the police force did not care about these losses because of the race of the victims. Many did not believe that the perpetrator of the crime would himself be Black. Some believe that police planted evidence against Williams so they could establish a fall man, or that the real crimes were perpetrated by the KKK.

We want to draw a few conclusions here to set up the discussion for this chapter. First, it is a common defense strategy to weave a narrative about police conspiracy.

This is an effective strategy because police officers and even whole departments *have* sometimes framed people (Romo 2018). We don't think these kinds of defenses are unreasonable. We think there is a difference between, on the one hand, considering that a conspiracy theory *might* be true when considering whether to convict a person who is presumed innocent in a court of law on the one hand, and on the other hand, full on *believing* a conspiracy for which there is scant or no evidence outside of that context.

The police cover-up conspiracies discussed above provide an interesting array of power dynamics, each of which provides us with insight into the potential response and the willingness to accept the theory. Steven Avery was a poor, white, rural man who had already been irreversibly and seriously harmed by the police department and the government. His story may have resonated powerfully with other members of his demographic—a demographic which often feels invisible, under-supported, and demonized. Simpson and Williams were both Black men, but from very different positions of power. These power dynamics no doubt affect public perception about their guilt and about possible conspiratorial behavior involved in their respective arrests and prosecutions.

Second, there may be cases in which heightened skepticism against certain institutions is warranted, and this may sometimes even result in acquittals of guilty people. We'll discuss this at greater length in the next chapter.

Conspiracy Theories and Systems of Power

Conspiracy theories are often used to cast a negative light on the actions of groups advocating for social change. We saw an example of this at the beginning of Chapter 11 when the response to the phenomenon of more women entering the workforce caused The Satanic Panic. As Martin Luther King Jr says powerfully in "Letter from a Birmingham Jail," "We know through painful experience that freedom is never voluntarily given by the oppressor; it must be demanded by the oppressed." The fact that he wrote these words on toilet paper in a jail cell highlights the fact that society at large is often none too fond of simply moving over and allowing oppressed groups a seat at the table. In this chapter, we'll look at another category of harms caused by the proliferation of conspiracy theories—harms that demonize justice efforts in order to maintain the status quo, along with harms that fall disproportionately on marginalized groups.

Exacerbating these problems is the issue of epistemic injustice. Some voices are not as apt to be listened to as others by the community as a whole. When members of certain populations try to explain what's taking place in their communities, they struggle to bring about change because of the power dynamics in play. Society is more inclined to listen to the voices of those who are powerful. As a result, it can be particularly difficult for members of minority populations to exercise a loud counter-response to a conspiratorial narrative.

"It Wasn't Me, It Was Antifa"

The label "Antifa" stands for "anti-fascist." This is the great irony of the set of conspiracy theories that give the group (if such a unified group exists at all) a starring role. Some people protesting neo-Nazism, white-supremacy, and far-right extremism might identify as being part of the "Antifa" movement, even though it is highly decentralized. It is also worth noting that, in many of these conversations, because of Antifa's association with the Black Lives Matter Movement, "Antifa" is used as something of a dog whistle to signify "black."

We want to briefly discuss conspiracy theories that target Antifa in this section because this loose collection of associated groups attempts to combat the kind of mindset that leads to belief in the kinds of conspiracies that are discussed below.

One conspiracy theory that enjoys wide support is that the attack on the Capitol on January 6th was not perpetrated by Trump supporters (Qiu 2022). Instead, it was a co-ordinated and elaborate effort by Antifa (and, on some accounts, the FBI) to *make it appear* that Trump supporters were behind it. Conspiracists believe that this is the case even though the rioters themselves are perfectly clear about their identities and motivations. Plenty of them have been tried and convicted, during which process, their motivations were laid bare and there was plenty of opportunity for the truth to come out through all the presentation of evidence (Popli and Zorthian 2022). Here, again, we see the unfalsifiability component of conspiracy theories on full display. Any evidence that suggests that the facially obvious Trump supporters who attacked the Capitol *were, in fact* Trump supporters is just further evidence to demonstrate just how deep the conspiracy runs and how realistic the crisis actors are. Strangely, this set of conspiracists seeks to blame a shadowy group standing for antifascism for one of the most frightening events supporting fascism in recent history.

There are at least two responses available to people who want to defend those present at the Capitol on January 6th 2021. One response is to deny the severity of the event. Plenty of politicians and ordinary citizens have adopted this approach. The other approach is to acknowledge how bad it was, but to insist that it wasn't Trump supporters at all, despite the complete lack of evidence for that position.

The Gay Agenda

For decades, opponents of homosexuality and gay marriage have been claiming that members of the LGBTQ community are out to destroy traditional marriage and dismantle the American way of life. Conspiratorial claims don't just take place in private chat groups on the Internet; they're spread by politicians, pastors, and shock jocks around the country. The dominant theme of these theories tends to be, well, essentially, that the LGBTQ community is on a constant singly focused mission to feminize men, turn women into lesbian radical feminists, and molest children in order to recruit them into the gay community.

The methods conspiracists believe are being used in order to bring these goals about tend to be as bizarre as any conspiracy theory we've discussed so far. Texas Congressman Louie Gohmert claimed that gay people are infiltrating the military in order to make our armed forces soft and weak. He claimed that they just "get massages all the time" making us "vulnerable to terrorism" (Brinker 2014). One common conspiracy is that perpetrators of the "gay agenda" are trying to create a new stage in human evolution, the "genderless person." They claim that the gay agenda wants to do away with traditional masculinity. According to many such theories, one of the main goals of the gay agenda is to bring down Christianity, because in many of its denominations, Christianity (or, at least, Christian leaders) advocate for traditional gender roles and norms, along with heterosexuality and celibacy until marriage, which they view as between a man and a woman. Destroy gender norms, and Christianity comes crashing down with it, or so these conspiracists believe. We can see that this set of conspiracy theories relies heavily on the Manichean idea of a battle between good and evil. Many of these conspiracists argue that advocates of "the gay agenda" are on the dark side of the battle between the darkness and the light. These conspiracists don't mince words about their commitment to this worldview, even going so far as to claim that homosexuality is the explanation for natural disasters—God is sending his destructive forces to punish humanity for its permissiveness (Pasha-Robinson 2017).

Belief in these kinds of conspiracy theories has significant consequences for members of the LGBTQ community. Conspiracists have insisted that the reason trans people want to use the bathroom that is appropriate for their gender identity is to molest the children who use that restroom. As a result, as of 2017, sixteen states have adopted laws prohibiting trans individuals from using the bathroom of their choice. This, despite the fact that there is no evidence for the claim that abuse against children perpetrated by trans people, in bathrooms or otherwise, happens regularly (or at all). That said, there *is* evidence to support the claim that when trans and non-binary individuals are restricted when it comes to the bathroom they are allowed to use, *they* are more vulnerable to sexual assault (Harvard School 2022).

This has not been the only legislation passed to fight the "gay agenda." There are laws that prohibit teaching safe sex when it comes to LGBTQ sexuality in sex ed classes because of the fear that advocates of the gay agenda, who conspiracists believe dominate the schools, are trying to indoctrinate our children in order to get them to reject their own gender identity and, ultimately, to bring down Christianity.

This set of conspiracy theories is tricky to handle *as* a set of *conspiracies*. Many would say, "I'm not a conspiracy theorist, I just believe in traditional sexualities and gender identities." The fact is, however, that insofar as one's "defense" of such a position is to postulate the existence of secret schemes for global domination for which one has no evidence, one *is* advocating for a conspiracy theory. We don't at all agree with such a position, but there is room in logical space for a view that says, "I believe homosexuality is wrong. I don't believe the members of the LGBTQ community are trying to convert my children or destroy religion and American institutions." That is, there is room to oppose gay rights, but not to do so for conspiratorial reasons.

Conspiracies that result in a restriction of rights for a particular segment of people are particularly troubling. If rights are ever going to be restricted, it should be for a defensible reason—a person should have compelling evidence behind their argument for it. For instance, if a child's brain is still developing until their early twenties, that may serve as a good argument for restricting their access to alcohol. Conspiracy theories do not count as good arguments for the restriction of rights. This causes real, significant harms. These kinds of policies result in increases in suicide and suicide attempts among members of the LGBTQ community (Haas et al., 2011). Conspiracy theories regarding the "gay agenda" put significant obstacles in the path of members of the LGBTQ population toward living authentic and happy lives.

Immigration and Islam

If there were a class titled "Demagoguery 101," the majority of the content would likely focus on fear-mongering, xenophobia, racism, creating and sustaining in-group/out-group dynamics, otherizing, demonizing, and repetition. Demagogues insist to their rapt audiences that the out-group is coming for them, that the out-group is different from them racially and culturally, and that the out-group fights on the side on the darkness. They play those psychological characteristics commonly found in conspiracists like a fiddle. Demagogues and their cohort repeat these rhetorical points over and over again. Hitler and his accomplices used these weapons to great effect in their campaign against Jewish persons and other groups that they deemed to be "undesirable" or inconsistent with their mission.

What was effective then is still effective now. There are many absurd conspiracy theories about immigrants. The dominant one, and the theory from which most of the others branch off, is "The Great Replacement" theory that we discussed in Chapter 10, which, you'll recall, claims that Democrats are intentionally bringing immigrants into the country in order to replace white people and, more dramatically and absurdly, initiate a white genocide. Consider the racist and antisemitic conspiracy theory perpetrated by Trump and his supporters that there was an enormous caravan of migrants, funded by—you guessed it—George Soros—headed for the United States border with the intention of invasion or attack . . . or something like that. It was unclear. Trump stoked fears about these migrants in the months leading up to the 2016 election. It turns out there was no attack, no invasion. It's a truism that migrants move— there's nothing inherently ominous about that; many of them are escaping violence, political unrest, poverty, or other dangerous conditions in the areas from which they came. The caravan stopped in Tijuana, but many of the people involved stopped elsewhere, at earlier points on the journey. Some no doubt tried to cross the border illegally, but others (the majority) applied for asylum and waited for their chance at safety (Alvarez 2019). There was no sign of Soros.

Then, of course, there are the conspiracies theories that focus on Muslims. These have always been present in American culture, but they gained in popularity and intensity after 9/11 and then again during the 2016 election. One of the dominant conspiracy theories in this camp is that Muslims are attempting to infiltrate the United States with the goal of imposing Sharia Law. Then-candidate Trump ran a campaign that vowed to prevent, or, at the very least, very strictly monitor Muslim immigration into the country. The same month that Trump was sworn into office, he issued an Executive Order banning immigration from seven predominantly Muslim countries (ACLU 2020).

We've discussed the use of conspiracy theories as a political tool in earlier chapters. Tragically, these kinds of strategies have even more force, and they are even more effective, when they capitalize on people's existing fears about race and cultural difference. This isn't that surprising when we look at the psychological mechanisms that fuel belief in conspiracy theories. People believe conspiracies, at least in part, to protect their own psychological well-being. If Muslims, immigrants, or Black people, are "the other," then, such people feel that *they, themselves* must be part of the in-group—the non-other! Conspiracy theories, especially those that are designed for political purposes, capitalize on people's fear. In the case of all of the groups we've discussed so far, it's that critical fear of the loss of one's own power and one's own identity. People are afraid that the culture with which they feel at home and comfortable will disappear, they're afraid of losing their job or their power, and they're afraid of no longer having a reliable set of cultural expectations. They can fight these losses by fighting the bogeymen presented in the form of conspiracy theories.

15
Conspiracy Theories as a Social Problem

As the World Burns, Part Two

In early 2021, The World Meteorological Society issued its report on climate, and the results were frightening. In 2020, the average global temperature was 1.2 degrees Celsius higher than pre-industrial levels. The scientific consensus on the topic is that we need to keep that number below 1.5 degrees higher (than preindustrial levels) to ward off the most concerning effects of climate change. We have already seen some of these consequences in the form of melting ice caps, coral bleaching, rising sea levels, and extreme weather events. On the topic of the results of the report, Secretary-General António Guterres of the United Nations said, "We are on the verge of the abyss."

We've discussed many social problems in this book: the fragility of our democracy, the strength of our public health system, the resiliency and dependability of our court systems in the face of hysteria. None of the problems hold a candle to the severity of the consequences of anthropogenic climate change. None of them.

Sometimes it doesn't take much to get a conspiracy theory going. In the mid 1990s the United States Air Force published a paper that considered the possibility of weather modification, presumably for the purposes of fighting climate change. Within weeks right wing radio hosts led by Art Bell (you didn't think that there would be an entire book on conspiracy theories that didn't mention Art Bell, did you?) had begun to theorize that the government was messing with the weather for nefarious purposes. Shortly thereafter the various theories about weather control had coalesced into the Chemtrails conspiracy theory.

The Chemtrails conspiracy theory was that many of the contrails seen in the wake of jet airplanes were actually chemtrails. Contrails are condensation trails (hence the name "contrails") produced by airplanes flying at certain high altitudes. Chemtrails are contrails that contain dangerous chemicals. These dangerous chemicals, according to the theories, are being placed into the atmosphere by the United States Government. On some versions of the conspiracy theory, the US government is doing this in collusion with other world governments.

What are the dangerous chemicals and what do they do? The theory is not entirely clear in response to the first question. They're just dangerous (unlike other conspiracy theories we've discussed, there is a real lack of detail in this one). The theory is clearer in response to the second question, but there's no real consensus among Chemtrails conspiracy theorists. Some maintain that the government is putting chemicals in the atmosphere to control our minds (or possibly our thoughts). Others maintain that the government is using chemicals to control the population either through forced sterilization or reduced sex drive. Still others claim that the chemicals are being used to reduce our lifespans. And, of course, there are still some who maintain that it is just about controlling the weather, even though nothing in the Air Force report made reference to chemtrails. Many take these theories a step further by postulating that the various government activities and experiments are the cause of climate change. Some go even further than that in maintaining that the government is using chemtrails to fight climate change (in a sort of make-shift geoengineering project) but are doing so covertly, as the chemicals being used have significant deleterious effects on health and other parts of the environment.

One point that has been made repeatedly in this book is that the evidence for conspiracy theories tends to be not very good. It usually comes in the form of some story of how the conspiracy could have been carried out, and then relies on the person to whom it is presented to accept it. This works because some people really want to believe that the conspiracies theories are true. While this holds across nearly all conspiracy theories, it is not the case that all such theories are created equal. In some cases, the story is more plausible than in others. In some cases, the evidence offered provides more justification for accepting a particular conspiracy theory and in some cases it provides less. Evidence for conspiracy theories exists on a continuum. At the weakest end of the continuum is where one finds the evidence for the Chemtrails conspiracy theory. No conspiracy theory contains less support (except perhaps for the Birds Aren't Real conspiracy theory).

Support for the Chemtrails conspiracy theory rests largely on the observation that some contrails look different from and last longer than the others (and the related claim that contrails used to look different—a fact that is easily disputed by literally thousands of photographs). The hastily drawn conclusion is that they must contain chemicals designed to control our minds, or whatever. Other bits of "evidence" for the existence of chemtrails are the occasional occurrence of large numbers of birds dying in a particular location over a short period of time and some video footage of a plane dumping fuel.

Each of these things is easily accounted for. The difference in duration and appearance of contrails lies in the fact that a variety of factors, such as weather and atmospheric pressure, will bear on the properties that any particular contrail has. Sometimes birds die in great numbers. The explanation for one instance of this happening (2000 geese were found dead in Idaho) was avian cholera (Insider 2019). Occasionally planes dump fuel. Typically, it's because the pilots need to make an emergency landing. Despite the fact that the Chemtrails conspiracy theory rests on the flimsiest evidence, and to succeed would require literally tens of thousands of government and airline employees to keep all aspects of it secret, it remains one of the more widely held conspiracy theories. Our suspicion is that it endures mostly because plenty of people just really love a good "the government is out to get us" story. Once again, confirmation bias rears its ugly head!

This conspiracy theory has existed for a long time and doesn't cause the kind of political unrest to which some of the others lead. It doesn't target traditionally marginalized populations or fuel hate mongering and otherizing. The dominant problem with this kind of theory, however, is that if you believe it, or, at the very least, if you believe some version of it, you will be less inclined to think that we need to act immediately to do something about climate change. You may believe that the government already has it handled—after all, they have more control over the skies than the average person is inclined to think. Making matters worse is the fact that this is far from the only conspiracy theory that touches directly on the question of climate change (recall The Great Reset Theory that we discussed in Chapter 10). Belief in conspiracy theories may be a barrier to taking the kind of group action that we need to take in order to make some difference to the severity of climate change.

Individual Responsibility?

Anyone on social media has probably had their fair share of experiences with people who regularly share conspiratorial content and postulate conspiracy theories in their Facebook statuses. This can be very frustrating, especially when the matter about which they are speculating has serious consequences for all sorts of stakeholders. That said, there has also been much discussion in this book about features that can and should mitigate our moral judgments about individual believers. In Chapter 12, we discussed belief voluntarism—the idea that we have full control over what we believe—and we provided reasons for thinking that belief voluntarism is false. That is, we provided reasons for thinking that we may not have full control over what it is that we believe, or at least we may not have control over all of it. If this is the case, then our uncle in his tin foil hat may have little control over whether he thinks that the Moon landing didn't really happen or that 9/11 was an inside job.

That said, though our uncle may not have control over what he believes, he may have a more meta-level control over the processes he employs when forming them. He knows, for instance, the number and variety of sources that he is accessing when attaining evidence. He knows whether he is willing to have good faith discussions with people with whom he disagrees. He knows whether he is willing to change his beliefs when presented with evidence that contradicts their content. If he is unwilling to question his conspiratorial attitudes or to employ any epistemic best practices at all when coming to beliefs, he may be morally blameworthy for being a conspiracy theorist, especially when the conspiracy theories he believes hurt other people.

Another mitigating factor for individual moral responsibility is deception. A person often has very little control over their social circle, and their beliefs and attitudes are often strongly influenced by the beliefs and attitudes of those around them and of the leaders and public figures that the people around them respect. For example, a son may respect his father, and his father may respect a politician. This could easily be true for all members of a given population. If a member of that population lies, those lies spread quickly through the group. Philosophers frequently say, "knowledge is the norm of assertion." As social creatures who use language to coordinate and work together, when someone in our trusted group makes a claim, we are inclined, immediately and often unreflectively, to believe that that person has evidence for the claim. When one of the people helping to spin that web of belief, in our example let's say it's the politician, lies, that lie soon becomes part of the more integrated system of beliefs held by the entire community, since they believe that a person whom they trust surely wouldn't make alarming assertions without evidence to back them up. This is just one way in which conspiracy theories, at their worst, are a social disease—they are remarkably contagious.

If a person is unknowingly deceived, this mitigates the extent to which they are blameworthy for what they believe. There may even be reason to feel sorry for them—we often understand people who have been lied to as victims of the deceiver, depending on the content and severity of the lie. That said, the old adage, "Fool me once, shame on you, fool me twice, shame on me" might be applicable here. If Tom lies to Jane once, it might be reasonable for her to continue to trust him. If Tom lies to Jane over and over again, then Jane may be at least partially culpable for not rethinking whether she should continue to trust Tom. Trust in the truthfulness of a source should be based upon a reliable disposition and track record on the part of that source as a truth-teller. If it becomes clear that the source is not a reliable, we shouldn't believe what they say, or we should be reluctant to do so in the absence of further investigation.

A person who believes something on the basis of deception may also be blameworthy if they put their trust in someone and then simply stop paying attention to how things turn out. If Tom donates five dollars toward a local community carnival and then never verifies that the carnival actually took place, for all he knows, someone just hoodwinked him out of his five dollars. On a more serious note, if voters vote a certain political candidate into office because that candidate pledged to do something about climate change and then that candidate gets voted into office and does nothing about climate change (perhaps their voting record actually even makes things worse), that should matter to the voters. If they simply vote, stop paying attention, and then vote for the same candidate again when they come up for re-election, the voter is at least partially responsible for the outcome. The degree of the responsibility may not be nearly as great as the degree of responsibility attributable to the politician, but it is not non-existent.

We have seen that individual believers of conspiracy theories are, under some circumstances, at least partially blameworthy for believing such theories and for the consequences that come about as a result of such belief. Other individual actors carry a greater degree of responsibility: the individuals creating and passing along the theories, especially if they are doing so for political reasons. These people aren't simply falling into certain common conspiratorial traps, they are *profiting or otherwise benefitting from the fact that others do!* These people frequently have extremely selfish motivations, preferring to confer some comparatively small benefit on themselves at the expense of serious harm to others.

Though individual actors bear some of the responsibility for belief in conspiracy theories, we propose that the most useful approach is to think about this as a social problem akin to other social problems.

Conspiracy Theories as Social Problems

In an ideal world, there would be no conspiracy theories. Or, maybe, in an ideal world, conspiracy theories would exist simply as entertainment or brain teasers. We've provided evidence in this book to suggest that conspiracy theories have nearly always existed. Some of them persist through the centuries, perhaps becoming even more entrenched as they go. Who knows whether Richard III ordered the murder of the Princes in the Tower, but everyone believes it now! (Except, of course, for the disquietingly passionate Richard III society). Conspiracy theories speak to and interact with basic facts about human psychology. There's no reason to think those facts are going to change. So there is good reason to think that conspiracy theories will always be with us, wreaking the same old havoc as always.

There's disagreement among metaethicists about what we should even be doing when we construct ethical theories. Should we be attempting to articulate a set of behaviors in which a person should engage in an ideal world? Are ethical principles a set of idealized standards? This kind of approach is frequently referred to by philosophers as "ideal theory." Should we instead accept the fact that the world is not perfect and never will be? Should we admit that conspiracies will always be with us and try to do the best that we can in light of that fact? This approach is called "non-ideal theory" and it is our preferred approach. As they say, "don't let the perfect be the enemy of the good!" By way of analogy, societies have struggled with addiction to drugs and alcohol and the corresponding consequences of those phenomena for quite some time. The same is true with prostitution. One approach we could take toward these problems is to go heavy on the attributions of individual moral responsibility. That may help us to carve out in-groups and out-groups—violators of and non-violators of social policy. That approach doesn't help much except with our egos.

Instead, as in the cases of drugs, alcohol, and prostitution, when it comes to belief in conspiracy theories, we think our approach should be similar to an approach we would take to any significant social problem—figure out how best to treat it, reduce the amount of it, and minimize the negative consequences of it. Our final discussion in this book will present some proposals for how this might be done.

Our first proposal is to **improve the level of trust that people have in public institutions.** We don't want to sound more optimistic about this than the situation warrants. Not all distrust in institutions happens because those institutions have done things to warrant distrust. For example, one way in which former President Trump sought to control the narrative surrounding his presidency was to convince his followers that the "mainstream media" was out to get him. He used the rhetorical device of repetition to reinforce the idea, over and over again, that those news network (and CNN in particular) who were reporting negative aspects of his administration were "fake news." As a result, we have many acquaintances (and we're sure you do too!) who believe that supporting one's point by referencing a CNN article is tantamount to referencing the *National Enquirer*.

Others distrust the media simply because it isn't satisfying those psychological needs for well-being that we discussed in Chapter 11—reports of deadly viruses, rogue cops, or riots don't make a person feel safe, secure, or happy. If the media is reporting events and behavior that paint your favored political candidate in a bad light, that may serve to undermine your sense that you have joined the right team or made the right choice. So, rather than give up one's social group or sense of identity, one distrusts the media. These cases are trickier and it is less clear what should be done about them.

There are cases in which sources of distrust are a little more obvious and are cases in which it might be possible to do more to improve the situation. Consider a couple of the cases described in the beginning of the last chapter: the Wayne Williams case and the O.J. Simpson case. We don't think that it is too blunt to say that conspiracy theories in these two cases were able to gain traction because plenty of members of the Black population did not trust the police. As recent events have demonstrated, many still don't, and for good reason. Police have been a source of violence against black bodies since the inception of the police force. Conspiracy theories about duplicitous police officers and against whole departments will continue to gain traction so long as tensions still exist and police continue to resist the kinds of reforms that might serve to build public trust.

The public health system presents a similar kind of case. It has not always been clear to people of color that the medical profession cares what happens to their bodies. Frankly there have been times when it's clear that they don't. The Tuskegee Syphilis Experiment that took place between the 1930s and the 1970s was such a case. The study was presented to the participants, all Black men from Tuskegee, as a treatment for syphilis, but the drugs they were given were placebos. The real purpose of the study was to see what the physical consequences were when syphilis was left untreated. Many of the participants suffered and died as a result of a perfectly curable and treatable disease.

We saw in Chapter 11 that, in terms of demographics, Black people and people of color are more likely to believe conspiracy theories. We found this to be in somewhat jarring contrast with some of the other demographics that dominate the category of conspiracy theory believers (Republican, rural, and so forth). A history of institutional behavior leading to distrust might go a long way toward explaining that. The only way out may be to express an unequivocal commitment both to listen to and acknowledge the concerns of these communities and the past bad actions committed by institutions and to take *real, concrete action* to prevent and redress this wrongdoing.

Our second proposal (which we recognize to be a long shot) is to **work toward a society that is more equitable in general.** We believe that this is supported by what we've learned about the psychology at play when it comes to conspiratorial belief. People believe conspiracy theories, in many instances, to feel good about themselves. This can often be because they are feeling misunderstood, disenfranchised, and feel that they are not important or powerful for one reason or another.

The demographics that we looked at in Chapter 11 support the idea that conspiracy theories, or, at least, some conspiracy theories, are popular with low-income individuals with a high school diploma or less who make under $50,000 a year. We think it would be useful to have a more robust portfolio of social services funded for all people so that they find themselves in a position to seek out different sources of power and self-respect. This is tricky, because those kinds of policies won't tend to be the kinds for which this demographic would vote, but we think this would be a good thing for everyone involved.

Making college education more affordable is one such potential solution. Earlier in the book, we mentioned that it would be naive to think that we could just offer increased access to critical thinking classes and it would solve the whole problem. That said, we don't think that increased access to high quality education could *hurt*. We think it would help; we just don't think it would fix the situation entirely. Critical thinking classes would be useful because many conspiracy theories rely on fallacious reasoning, as we've seen examples of on multiple occasions throughout this book. Noteworthy among them are the false cause fallacy (the mistaken belief that because one event preceded another, that event caused the other), the wishful thinking fallacy (the mistaken belief that wanting something to be true is sufficient for making that thing true), the denial fallacy (the mistaken belief that wanting something to be false is sufficient for making that thing false), and the fallacious appeal to authority (appealing to the authority of someone who is not an expert or is an expert in the wrong subject).

More importantly, however, access to advanced education can help people to achieve more control over the nature and course of their lives. It gives them access to higher paying jobs. All of this is contingent of course, on the idea that school does not trap people in debt, so the education must genuinely be affordable, not only "affordable" by taking out interest bearing loans which a person may never be able to afford to pay off.

In addition to education, our proposed solution is that other social services are available to help people live happy, healthy, productive lives so that they're less inclined to demonize others or try to find scapegoats for their problems. We think that conspiracy theories are the consequences of an unrestrained capitalism that does not set aside money to solve these kinds of social problems of disenfranchisement.

Our third proposal has to do with **communication of various forms.** People doing other kinds of work feel disconnected from people who are employed doing research of various sorts. People are strongly divided into groups by both socioeconomic class and profession. As a result, the human element of interaction is missing. When people don't see one-another as human beings, it is easier for them to discount or demonize one another.

As we noted in Chapter 11, one common feature of conspiracies theorists is that they don't trust people whom they characterize as "elites." We can go in one of two directions with this (or, perhaps we can employ a little of both). First, we can acknowledge that people don't care for "elites" and we can find way of designing important messages that fully acknowledge that fact. It helps to have communication partners in various communities to help transmit information in ways that is friendly and effective.

Second, we can make the work of researchers and other professionals more accessible to people. We can have more publications and educational programs of various sorts designed for a wide audience. Universities can and should start to think of the job responsibilities of at least some of their faculty members as communicating regularly with the public in various ways. This should not be some subordinate duty, relegated to faculty they don't trust with "more important" responsibilities; it should be viewed as one of a faculty member's most pressing, useful, and meaningful obligations. In communications with community members, academics and researchers should be approachable and accessible and willing to discuss important topics of the day in the clearest and most concise ways possible. This in no way means abandoning rigor in research, it simply means taking on communication as an important social role, especially in fields that are the most impactful for the problems that we face.

Our final proposal is that we **think about the ways that we want our community to be able to transmit information.** New communication technology has sped along largely unrestricted. The same thing has happened with news media. We should have open discussions about whether our web of belief should be so easily intertwined with poisonous deception and nonsense. Just because a particular technology is possible, doesn't mean it is advisable. This is advice that is true about parks of resurrected dinosaurs, and also true about some methods of communication. It is at least possible that philosophers like Mill who seemed to advocate for complete unrestricted free speech could not have anticipated in any way the toxic circumstances in which we find ourselves. We should at least be open to conversations about how far we want this technology to go, and about whether it has already gone too far.

Conspiracy theories will always be with us. There may be little we can do about it, and the situation may only intensify. The first step is to acknowledge the magnitude of the problem. The second may be to recognize that our major social problems are interconnected.

References

ACLU. 2020. Timeline of the Muslim Ban <www.aclu-wa.org/pages/timeline-muslim-ban>.

ADL. 2021. 'The Great Replacement': An Explainer. <www.adl.org/resources/backgrounders/the-great-replacement-an-explainer>.

Alvarez, Priscilla. 2019. What Happened to the Migrant Caravans? CNN Politics <www.cnn.com/2019/03/04/politics/migrant-caravans-trump-immigration/index.html>.

American Association for the Advancement of Science. 2021. Blind Trust in Social Media Cements Conspiracy Beliefs. EurekAlert! (March) <www.eurekalert.org/news-releases/877317>.

Bažant, Zden k P., and Mathieu Verdure. 2007. Mechanics of Progressive Collapse: Learning from World Trade Center and Building Demolitions. *Journal of Engineering Mechanics* 133:3 (March).

BBC News. 2020. Singapore Approves Lab-Grown 'Chicken' Meat. <www.bbc.com/news/business-55155741>.

BBC News. 2021. Twitter Suspends 70,000 Accounts Linked to QAnon. <www.bbc.com/news/technology-55638558>.

Brennan Center for Justice. 2021. Voting Laws Roundup: July 2021. 2021. <www.brennancenter.org/our-work/research-reports/voting-laws-roundup-july-2021>.

Brinker, Luke. 2014. Louie Gohmert Worries that Gay Soldiers 'Getting Massages All Day' Leave Us 'Vulnerable to Terrorism'. *Salon* (October 24th) <www.salon.com/2014/10/24/louie_gohmert_worries_that_g

ay_soldiers_getting_massages_all_day_leave_us_vulnerable_to_terrorism>.

Brittanica. Illuminati: Designations for Various Groups <www.britannica.com/topic/illuminati-group-designation>.

Camus, Albert. 2018 [1942]. *The Myth of Sisyphus*. New York: Vintage.

Cascone, Sarah. 2020. Archeologists May Have Finally Solved the Mystery of the Disappearance of Roanoke's Lost Colony. *Artnet News*. <https://news.artnet.com/art-world/archaeologists-mystery-lost-roanoke-lost-colony-1921594>.

Cassam, Quassim. 2019. *Conspiracy Theories*. Polity.

Cheathem, Mark R. 2019. Conspiracy Theories Abounded in 19th-Century Politics <www.smithsonianmag.com/history/conspiracy-theories-abounded-19th-century-american-politics-180971940>.

Clifford, W.K. 1999 [1877]. The Ethics of Belief. In T. Madigan, ed., *The Ethics of Belief and Other Essays*. Amherst: Prometheus.

Coady, David, ed. 2018 [2006]. *Conspiracy Theories: The Philosophical Debate*. Routledge.

———. 2018b. An Introduction to the Philosophical Debate about Conspiracy Theories. In Coady 2018a.

The Conversation. 2021. Support for QAnon Is Hard to Measure—and Polls May Overestimate It. <https://theconversation.com/support-for-qanon-is-hard-to-measure-and-polls-may-overestimate-it-156020>.

Coppenger Brett, and Joshua Heter. 2020. How to Build a Conspiracy Theory. In Greene and Robison-Greene 2020.

Cybersecurity and Infrastructure Security Agency. 2020. Joint Statement. <www.cisa.gov/news/2020/11/12/joint-statement-elections-infrastructure-government-coordinating-council-election>.

Dando-Collins, Stephen. 2010. *The Great Fire of Rome: The Fall of the Emperor Nero and His City*. Da Capo.

Däniken, Erich von. 1999 [1968]. *Chariots of the Gods: Unsolved Mysteries of the Past*. New York: Berkley Books.

Dentith, M R.X. 2020. From Alien Shape-Shifting Lizards to the Dodgy Dossier. In Greene and Robison-Greene 2020.

Descartes, René. 1999. *Discourse on Method and Meditations on First Philosophy*. Indianapolis: Hackett.

Dio Cassius. 1916–1927. *Roman History*. Nine Volumes. Loeb Classical Library.

References

Douglas, K.M., R.M. Sutton, and A. Cichocka. 2017. The Psychology of Conspiracy Theories. *Current Directions in Psychological Science* 26:6.

Drinkwater, K., N. Dagnall, and A. Parker. 2012. Reality Testing, Conspiracy Theories, and Paranormal Beliefs. *Journal of Parapsychology* 76:1.

Elder, John. 2021. Conspiracy Theorists Lack Critical Thinking Skills: New Study. *The New Daily* (July 25th) <https://the-newdaily.com.au/life/science/2021/07/25/conspiracy-theo-rists-lack-critical-thinking>.

Ellis, Emma Grey. 2019. 'Epstein Didn't Kill Himself' and the Meme-ing of Conspiracy. *Wired*. (November) <www.wired.com/story/epstein-didnt-kill-himself-conspir-acy>.

Fast Company. 2018. Buckle Up! Here's a Timeline of George Soros Conspiracy Theories. <www.fastcompany.com/90247335/a-timeline-of-george-soros-conspiracy-theories>.

Gettier, Edmund L. 1963. Is Justified True Belief Knowledge? *Analysis* 23 (June).

Greene, Richard, and Rachel Robison-Greene, eds., *Conspiracy Theories: Philosophers Connect the Dots*. Chicago: Open Court.

Gregory, Matt. 2021. Verify: Donald Trump Has Raised 200 Million Dollars Since Election Day. Where Does It Go? *WUSA*. <www.wusa9.com/article/news/verify/donald-trump-has-raised-200-million-dollars-where-will-it-go/65-0b5c323d-e6d2-4ac4-827f-2fac647ea53d>.

Haas, Ann P., et al. 2011. Suicide and Suicide Risk in Lesbian, Gay, Bixesual, and Transgender Populations: Review and Recommendations. *Journal of Homosexuality* 58:1 (January) <https://www.ncbi.nlm.nih.gov/pmc/articles/PMC3662085>,

Harvard School of Public Health. 2022. Transgender Teens with Restricted Bathroom Access at Higher Risk of Sexual Assault. <www.hsph.harvard.edu/news/hsph-in-the-news/transgender-teens-restricted-bathroom-access-sex-ual-assault>.

Haslanger, S. and Kurtz M. 2006. *Persistence: Contemporary Readings*. MIT Press.

Hume, David. 1999. *Writings on Religion*. Open Court.

———. 2007. *Hume: An Enquiry Concerning Human Understanding and Other Writings*. Cambridge: Cambridge University Press.

References

Hutson, Matthew. 2017. Why It's So Hard to Keep a Secret. *Scientific American* (May) <www.scientificamerican.com/article/why-it-rsquo-s-so-hard-to-keep-a-secret>.

Insider. 2019. Some People in Idaho Say the Government Is Poisoning Them with Chemicals. <https://www.insider.com/popular-conspiracy-theories-united-states-2019-5#some-people-in-idaho-say-the-government-is-poisoning-them-with-chemicals-19.

Jewish Telegraphic Agency. 1992. U.S. Congressman Warns Hungary about Antisemitic Politician. <www.jta.org/1992/09/03/archive/u-s-congressman-warns-hungary-about-anti-semitic-politician>.

Josephus. 2001. *Jewish Antiquities*. Five volumes. Loeb Classical Library.

Marist Poll. 2021. NPR/PBS NewsHour/Marist National Poll: Trust in Elections, Threat to Democracy, November 2021. <https://maristpoll.marist.edu/polls/npr-pbs-newshour-marist-national-poll-trust-in-elections-threat-to-democracy-biden-approval-november-2021>.

McGreal, Scott A. 2021. Does Social Media Foster COVID-19 Conspiracy Theories? <www.psychologytoday.com/us/blog/unique-everybody-else/202112/does-social-media-foster-covid-19-conspiracy-theories>.

Oliver, J. Eric, and Thomas J. Wood. 2014. Conspiracy Theories and the Paranoid Style(s) of Mass Opinion. *American Journal of Political Science* 58:4. <www.jstor.org/stable/24363536>.

Pasha-Robinson, Lucy. 2017. Gay People to Blame for Hurricane Harvey, Say Evangelical Christian Leaders. *Independent* (6th September) <www.independent.co.uk/news/world/americas/gay-people-hurricane-harvey-blame-christian-leaders-texas-flooding-homosexuals-lgbt-a7933026.html>.

Pigden, Charles. 2020. Everyone's a Conspiracy Theorist. In Greene and Robison-Greene 2020.

Plato. 1997. *Plato: Complete Works*. Hackett.

Popli, Nik, and Julia Zorthian. 2022. What Happened to Jan. 6 Insurrectionists Arrested in the Year Since the Capitol Riot. *Time* <https://time.com/6133336/jan-6-capitol-riot-arrests-sentences>.

References

Price, Joe. 2021. TikTok Users Attempt to Push Conspiracy Theory Claiming the Snow in Texas Is Fake. <https://sports.yahoo.com/tiktok-users-attempt-push-conspiracy-213311075.html>.

Prooijen, Jan-Willem and Karen M. Douglas. 2017. <www.ncbi.nlm.nih.gov/pmc/articles/PMC5646574>.

PRRI. 2021. Understanding QAnon's Connection to American Politics, Religion, and Media Consumption. PRRI-IFYC (May 27th) <www.prri.org/research/qanon-conspiracy-american-politics–report>.

Qiu, Linda. 2022. A Year after the Breach, Falsehoods about Jan. 6 Persist. *New York Times* (January 5th) <www.nytimes.com/live/2022/01/05/us/jan-6-fact-check>.

Raymond, Adam K. 2016. The 70 Greatest Conspiracy Theories in Pop-Culture History. Vulture. <www.vulture.com/2016/10/pop-culture-conspiracy-theories-c-v-r.html>.

Reuters. 2020. Fact Check: Photograph Shows Nazi Bookkeeper, Not George Soros. <www.reuters.com/article/uk-factcheck-photo-soros-nazi-bookkeeper/fact-check-photographshows-nazi-bookkeeper-notgeorge-soros-idUSKBN23P2TP>.

Romo, Vanessa. 2018 Ex-Florida Police Chief Sentenced to Three Years for Framing Black Men and Teen. NPR WBEZ Chicago <www.npr.org/2018/11/28/671716640/ex-florida-police-chief-sentenced-to-3-years-for-framing-black-men-and-teen>.

Sadeghi,McKenzie. 2022. Fact Check: The COVID-19 Pandemic Is Not a Hoax. USA Today (January 5th) <www.usatoday.com/story/news/factcheck/2022/01/05/fact-check-post-faslely-claims-covid-19-pandemic-isnt-real/9076551002>.

Stekula, Dominik A., and Mark Pickup. 2021. Social Media, Cognitive Reflection, and Conspiracy Beliefs. *Frontiers in Political Science* <www.frontiersin.org/articles/10.3389/fpos.2021.647957/full#fn1>.

Suetonius. 2007. *The Twelve Caesars*. London: Penguin.

Swami, Viren, Rebecca Coles, Stefan Stieger, and Jakob Pietschnig. 2011. Conspiracist Ideation in Britain and Austria: Evidence of a Monological Belief System and Associations between Individual Psychological Differences and Real-world and Fictitious Conspiracy Theories. *British Journal of Psychology* 102:3.

Tacitus. 1948. *The Agricola and the Germania*. Translated by H. Mattingly. London: Penguin.

———. 1952. *The Annals and the Histories*. Chicago: Encyclopaedia Britannica.

Terry, Maury. 1988. *The Ultimate Evil: An Investigation into a Dangerous Satanic Cult*. London: Grafton.

Time. 2003. Conspiracy Theories <http://content.time.com/time/specials/packages/article/0,2 8804,1860871_1860876_1861003,00.html>.

United Nations. 2021. World on the Verge of Climate 'Abyss', as Temperature Rise Continues: UN Chief. UN News <https://news.un.org/en/story/2021/04/1090072>.

Vibert, Kevin (@VibertKevin). 2021. Just heard a conspiracy theory that the Loch Ness Monster is actually the ghost of an ancient dinosaur, and since it affects nothing and nobody, I've decided I believe it, as a treat for me. Twitter. 8:39 am June 7th 2021.

Weigel, Chris. 2014. Quotidian Confabulations: An Ethical Quandary Concerning Flashbulb Memories. *Theoretical and Applied Ethics* 3:1.

Wikipedia. Stoneman Douglas High School Shooting. <https://en.wikipedia.org/wiki/Stoneman_Douglas_High_S chool_shooting#Behavioral_issues_and_social_media>.

Williams, Bernard. 1973a. *Problems of the Self*. Cambridge University Press.

———. 1973b. The Makropulos Case: Reflections on the Tedium of Immortality. In Williams 1973a.

Wittgenstein, Ludwig. 1922. *Tractatus Logico-Philosophicus*. London: Routledge.

———. 1953. *Philosophical Investigations*. Oxford: Blackwell.

Yuhas, Alan. 2021. It's Time to Revisit the Satanic Panic. *New York Times* (March 31st). <https://www.nytimes.com/2021/03/31/us/satanic-panic.html>.

Index